Advanced Math   Core 3 for AQA

Welcome to Advanced Maths Essentials: Core 3 for AQA. This book will help you to improve your examination performance by focusing on the essential skills you will need in your AQA Core 3 examination. It has been divided by chapter into the main topics that need to be studied. Each chapter has then been divided by sub-headings, and the description below each sub-heading gives the AQA specification for that aspect of the topic.

The book contains scores of worked examples, each with clearly set-out steps to help solve the problem. You can then apply the steps to solve the Skills Check questions in the book and past exam questions at the end of each chapter. If you feel you need extra practice on any topic, you can try the Skills Check Extra exercises on the accompanying CD-ROM. At the back of this book there is a sample exam-style paper to help you test yourself before the big day.

Some of the questions in the book have a ⊙ symbol next to them. These questions have a PowerPoint® solution (on the CD-ROM) that guides you through suggested steps in solving the problem and setting out your answer clearly.

Using the CD-ROM

To use the accompanying CD-ROM simply put the disc in your CD-ROM drive, and the menu should appear automatically. If it doesn't automatically run on your PC:

1. Select the My Computer icon on your desktop.
2. Select the CD-ROM drive icon.
3. Select Open.
4. Select core3_for _aqa.exe.

If you don't have PowerPoint® on your computer you can download PowerPoint 2003 Viewer®. This will allow you to view and print the presentations. Download the viewer from http://www.microsoft.com

Pearson Education Limited
Edinburgh Gate
Harlow
Essex
CM20 2JE
England
www.longman.co.uk

First published 2006
Third impression 2010
ISBN 978-0-582-83683-9

Design by Ken Vail Graphic Design

Cover design by Raven Design

Typeset by Tech-Set, Gateshead

Printed in China EPC/03

The publisher's policy is to use paper manufactured from sustainable forests.

The publisher wishes to draw attention to the Single-User Licence Agreement at the back of the book.
Please read this agreement carefully before installing and using the CD-ROM.

The Publisher and Authors would like to thank Rosemary Smith for her significant contributions to Chapters 1 and 6 of this book.

We are grateful for permission from AQA to reproduce past exam questions. All such questions have a reference in the margin. AQA can accept no responsibility whatsoever for accuracy of any solutions or answers to these questions.

Every effort has been made to ensure that the structure and level of sample question papers matches the current specification requirements and that solutions are accurate. However, the publisher can accept no responsibility whatsoever for accuracy of any solutions or answers to these questions. Any such solutions or answers may not necessarily constitute all possible solutions.

1 Algebra and functions

1.1 Functions: domain and range

Definition of a function. Domain and range of a function.

First consider a **mapping**, which is a relationship between two sets of data. The set of elements being mapped is called the **domain** and the resulting set is called the **range**.

For example, consider the domain consisting of the set {4, 3, 2, 1} being mapped to the range consisting of the set {8, 6, 4, 2}.

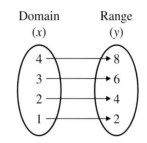

If the elements in the domain are represented by x and the elements in the range by y, the relationship is $y = 2x$.

> **Note:**
> A mapping may be described in words or by using algebra and may be represented by a graph.

There are four types of mapping:

- **one–one**, where each element in the domain is mapped to just one element in the range
- **many–one**, where more than one element in the domain can be mapped to an element in the range
- **one–many**, where an element in the domain can be mapped to more than one element in the range
- **many–many**, where more than one element in the domain can be mapped to more than one element in the range.

> **Note:**
> The expression one–one is read as one-to-one.

A **function** is a special mapping which satisfies both of the following conditions:

- it is defined for all elements of the domain;
- it is either **one–one** or **many–one**.

> **Note:**
> Mappings that are one–many or many–many are **not** functions.

Here are two examples of **one–one functions**:

a $f(x) = 3 - x$ for domain {3, 4, 5, 6}.

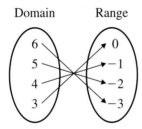

Range is {0, −1, −2, −3}

> **Note:**
> In **a**, the domain is a finite set of values.

b $f(x) = 2x - 1$ for all real values of x.

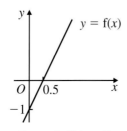

Range is $f(x) \in \mathbb{R}$

> **Note:**
> In **b**, the function can be written $f : x \mapsto 2x - 1, x \in \mathbb{R}$, where $x \in \mathbb{R}$ means x is real.

> **Note:**
> In **b**, the domain is a continuous interval.

Here are two examples of **many–one functions**:

a $f(x) = x^4$ for domain $\{-2, -1, 0, 1, 2\}$.

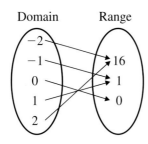

Range is {0, 1, 16}

b $f(x) = x^2 + 1$ for all real values of x.

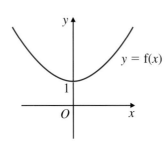

Range is $f(x) \geqslant 1$

Note:
When a graph can be drawn of $y = f(x)$, as in **b**, it is easy to see the domain and range: the domain is the set of all possible x-values and the range is the set of all possible y-values.

Example 1.1 State, with a reason, whether each of the following diagrams represents a function, for real values of x. If it is a function, state its range.

a $x^2 + y^2 = 25$

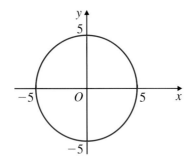

b $y = 4 - x^2$

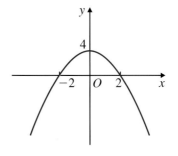

Step 1: Check the conditions for a mapping to be a function and make your conclusion.

a The mapping $x^2 + y^2 = 25$ is many–many, since, for example, when $x = 3$, $y = \pm 4$ and also when $x = -3$, $y = \pm 4$.

Also the mapping is not defined for all elements of the domain, since, for example, when $x = 6$, $y^2 = 25 - 36 = -9$, which has no real solutions.

Hence $x^2 + y^2 = 25$, for all real values of x, is not a function.

Recall:
Many–many mappings are not functions.

Step 2: If it is a function, use the sketch to state its range.

b The mapping $y = 4 - x^2$ is many–one and is defined for all real numbers, i.e. for all elements of the domain.

So $y = 4 - x^2$, for all real values of x, is a function.

The range is the set of real numbers y such that $y \leqslant 4$.

Tip:
Be careful with the use of $<$ and $\leqslant$. In **b**, when $x = 0$, $y = 4$, so the range is $y \leqslant 4$.

Example 1.2 The function f is such that $f(x) = x^2 - 4x - 5$, for all real values of x.

a The diagram shows a sketch of $y = f(x)$.

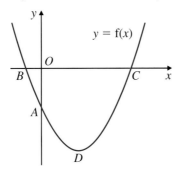

i Find the coordinates of A, B and C.

ii Write $f(x)$ in the form $(x - a)^2 + b$ and hence, or otherwise, find the coordinates of the minimum point D.

b State the range of f.

Step 1: Find the intercepts with the axes.

a i Consider $y = x^2 - 4x - 5$:

When $x = 0$, $y = -5$

When $y = 0$, $x^2 - 4x - 5 = 0$

$\Rightarrow \qquad (x - 5)(x + 1) = 0$

$\Rightarrow x = 5, x = -1$

A is the point $(0, -5)$, B is the point $(-1, 0)$ and C is the point $(5, 0)$.

Step 2: Complete the square to find a and b.

ii $f(x) = x^2 - 4x - 5$

$\qquad = (x - 2)^2 - 4 - 5$

$\qquad = (x - 2)^2 - 9$

Step 3: Write down the coordinates of the vertex of the curve.

So D has coordinates $(2, -9)$.

Step 4: Use the sketch and the coordinates of D to write down the range.

b The range of f is $f(x) \geqslant -9$.

> **Note:**
> You could use differentiation to find the coordinates of D (C1 Section 3.6).

> **Recall:**
> $y = (x - a)^2 + b$ has a minimum point at (a, b) (C1 Section 1.5).

> **Tip:**
> Look at the graph to see the set of all values that y can take. This is the range.

1.2 Composition of functions

Composition of functions.

Two or more functions may be combined to form a **composite function**. In doing this, you are finding the **composition** of two functions.

The notation fg is used to denote the composite function, where g is applied first, then f is applied to the result.

> **Note:**
> For the composite function fg to exist, the range of g must be a subset of the domain of f.

Example 1.3 The functions f and g are defined for real values of x as follows:

$$f(x) = 2x + 3 \qquad g(x) = x^2 - 2$$

a Find **i** $fg(x)$ **ii** $gf(5)$ **iii** $ff(x)$

b Solve $gf(x) = -1$.

Step 1: Apply g, then apply f to the result.

a i $\quad fg(x) = f(x^2 - 2)$
$\qquad\qquad = 2(x^2 - 2) + 3$

Step 2: Simplify.

$\qquad\qquad = 2x^2 - 1$

<table>
<tr><td>

Step 3: Substitute the given value into f, then apply g to the result.

Step 4: Evaluate.

</td></tr>
</table>

ii $\quad gf(5) = g(2 \times 5 + 3)$
$\qquad\qquad = g(13)$
$\qquad\qquad = 13^2 - 2$
$\qquad\qquad = 167$

Step 5: Apply f, then apply f again to the result.

iii $\quad ff(x) = f(2x + 3)$
$\qquad\qquad = 2(2x + 3) + 3$

Step 6: Simplify.

$\qquad\qquad = 4x + 9$

Step 1: Apply f, then apply g to the result.

b $\quad gf(x) = g(2x + 3)$
$\qquad\qquad = (2x + 3)^2 - 2$

Step 2: Simplify.

$\qquad\qquad = 4x^2 + 12x + 9 - 2$
$\qquad\qquad = 4x^2 + 12x + 7$

Step 3: Set up an equation and solve for x.

$\qquad\qquad\qquad gf(x) = -1$
$\Rightarrow \quad 4x^2 + 12x + 7 = -1$
$\qquad\quad 4x^2 + 12x + 8 = 0$
$\qquad\qquad\; x^2 + 3x + 2 = 0$
$\qquad\;\; (x + 1)(x + 2) = 0$
$\qquad x = -1, x = -2$

> **Tip:**
> Remember that
> $fg(x) \neq f(x) \times g(x)$.

> **Note:**
> $ff(x)$ is sometimes written $f^2(x)$.
> Remember that
> $ff(x) \neq f(x) \times f(x)$.

> **Note:**
> In general $fg(x) \neq gf(x)$.

> **Note:**
> This doesn't mean find $gf(-1)$.

> **Tip:**
> If the quadratic expression doesn't factorise, use the quadratic formula to solve the equation.

Example 1.4 For all real values of x, functions f and g are defined as follows:

$$f(x) = x^2 \qquad g(x) = 3x - 2$$

a Find $fg(x)$ and state the range of fg.
b Find $gf(x)$ and state the range of gf.
c Find the value of a for which $fg(a) = gf(a)$.

Step 1: Apply fg, i.e. apply g then apply f to the result.

a $\quad fg(x) = f(3x - 2)$
$\qquad\qquad = (3x - 2)^2$

Step 2: State the range of fg.

$\quad$ Since $(3x - 2)^2 \geqslant 0$, the range is $fg(x) \geqslant 0$.

Step 3: Apply gf, i.e. apply f, then apply g to the result.

b $\quad gf(x) = g(x^2)$
$\qquad\qquad = 3x^2 - 2$

Step 4: State the range of gf.

$\quad$ Since $3x^2 \geqslant 0$ for all real x, $3x^2 - 2 \geqslant -2$, so the range is $gf(x) \geqslant -2$.

Step 5: Substitute the given value into fg and gf, using your answers from **a** and **b**.

c $\quad fg(a) = (3a - 2)^2$
$\qquad\qquad = 9a^2 - 12a + 4$
$\quad gf(a) = 3a^2 - 2$

Step 6: Set up an equation and solve for a.

$$fg(a) = gf(a)$$
$\Rightarrow \qquad 9a^2 - 12a + 4 = 3a^2 - 2$
$\qquad\qquad 6a^2 - 12a + 6 = 0$
$(\div 6) \qquad\; a^2 - 2a + 1 = 0$
$\qquad\quad (a - 1)(a - 1) = 0$
$\qquad\qquad\qquad\; a = 1$

> **Tip:**
> In **a**, use the fact that this is in completed square form so you know its minimum (C1 Section 2.5). If you are unsure, do a sketch. A sketch may help in **b** also.

a **b**

> **Tip:**
> You could substitute your answer into fg and gf to check that you get the same result.

Inverse functions and their graphs.

The **inverse function**, f^{-1} represents the reverse mapping of the function f.

You must learn the following properties:

- For a function to have an inverse, it must be one–one.
- The domain of f^{-1} is the range of f.
- The range of f^{-1} is the domain of f.

For example the function f, where $f(x) = \frac{1}{2}x + 1$, with domain $\{3, 2, 1, 0\}$, has range $\{2.5, 2, 1.5, 1\}$. The inverse function sends each element from the range of f back to its original value in the domain of f, so under $f^{-1} : 2.5 \mapsto 3, 2 \mapsto 2, 1.5 \mapsto 1$ and $1 \mapsto 0$.

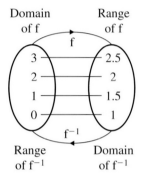

The domain of f^{-1} is $\{2.5, 2, 1.5, 1\}$ and the range of f^{-1} is $\{3, 2, 1, 0\}$.

Now extend the domain so that $f(x) = \frac{1}{2}x + 1$ for *all* real values of x.

The function f can be illustrated on a graph by the line $y = \frac{1}{2}x + 1$.

To show the inverse function f^{-1}, use the fact that the graphs of a function and its inverse are reflections of each other in the line $y = x$.

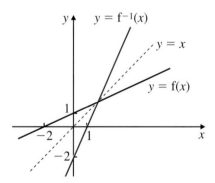

So if (x, y) lies on $y = f(x)$, then (y, x) lies on $y = f^{-1}(x)$.

To find the inverse function f^{-1}, let

$$y = \tfrac{1}{2}x + 1$$

Then interchange x and y:

$$x = \tfrac{1}{2}y + 1$$

Now make y the subject:

$$\tfrac{1}{2}y = x - 1$$
$$y = 2x - 2$$

Write this in terms of $f^{-1}(x)$:

$$f^{-1}(x) = 2x - 2, \quad x \in \mathbb{R}$$

Note:
The notation $f^{-1}(x)$ is easy to confuse with $f'(x)$ and $(f(x))^{-1}$. Remember that $f^{-1}(x)$ is the inverse function, $f'(x)$ is the derivative and $(f(x))^{-1}$ is the reciprocal function $\dfrac{1}{f(x)}$.

Tip:
To show this on a diagram, the scales on both axes must be the same.

Note:
In this example, the inverse function can be illustrated by a line.

Note:
You can interchange x and y at the beginning of the working or at the end.

Note:
This is the equation of the line $y = f^{-1}(x)$ in the diagram above.

Note:
Always write $f^{-1}(x)$ as a function of x.

If a function is many–one, by restricting its domain it can be made into a one–one function, enabling an inverse function to be found.

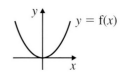

Consider the function f defined for all real values of x by $f(x) = x^2$. This is a many–one function since, for example, $f(-3) = (-3)^2 = 9$ and $f(3) = 3^2 = 9$.

If the domain is restricted to $x \geqslant 0$, then f is a one–one function and its inverse f^{-1} can be found, where $f^{-1}(x) = \sqrt{x}$, i.e. the positive square root of x.

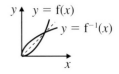

Example 1.5 The function f has domain $3 \leqslant x \leqslant 9$ and is defined by $f(x) = \dfrac{8}{1-x} + 6$.
A sketch of $y = f(x)$ is shown below.

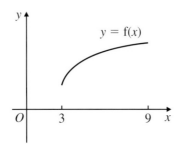

a Calculate $f(3)$ and $f(9)$.

b Find the range of f.

c The inverse function is f^{-1}.
Find f^{-1}, stating its domain and range.

d On the same set of axes, sketch $y = f(x)$ and $y = f^{-1}(x)$.

Step 1: Substitute appropriate x-values into the expression for f.

a $f(3) = \dfrac{8}{1-3} + 6 = 2$ and $f(9) = \dfrac{8}{1-9} + 6 = 5$

Step 2: Use the values obtained in **a** and check the sketch.

b The range of f is $2 \leqslant f(x) \leqslant 5$.

> **Note:**
> You also need to check the sketch to make sure there are no turning points.

Step 3: Let $y = f(x)$, then interchange x and y.

c Let $y = \dfrac{8}{1-x} + 6$.

Interchanging x and y gives

$$x = \frac{8}{1-y} + 6$$

Step 4: Make y the subject.

$$x - 6 = \frac{8}{1-y}$$

$$1 - y = \frac{8}{x-6}$$

$$y = 1 - \frac{8}{x-6}$$

Step 5: Use the relationship between f and f^{-1} to state the domain and range.

So $f^{-1}(x) = 1 - \dfrac{8}{x-6}$.

The domain of f^{-1} is $2 \leqslant x \leqslant 5$ and the range is $3 \leqslant f^{-1}(x) \leqslant 9$.

Step 6: Reflect the given curve in the line $y = x$ for the appropriate domain and range.

d

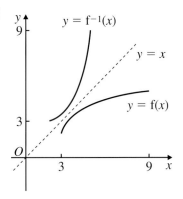

As a check, substitute a value for x in the function f, for example:

$$f(3) = \frac{8}{1-3} + 6 = 2$$

Now substitute the result into the inverse function f^{-1}:

$$f^{-1}(2) = 1 - \frac{8}{2-6} = 3$$

So, because the inverse function reverses the process of the original function,

$$f^{-1}f(x) = x$$

It is also true that

$$ff^{-1}(x) = x$$

This property can be used to **solve equations**.

Note:
f maps 3 to 2 and f^{-1} maps 2 back to 3.

Example 1.6 The function f has domain $x \geq -2$ and is defined by $f(x) = \sqrt{x + 2}$.

    a Find the inverse function, f^{-1}, stating its domain and range.

    b Find the value of x for which $f(x) = \frac{1}{2}$.

    c On the same set of axes, sketch $y = f(x)$ and $y = f^{-1}(x)$.

    d Solve $f(x) = f^{-1}(x)$.

Step 1: Let $y = f(x)$, then interchange x and y.

a Let $y = \sqrt{x + 2}$.

Interchanging x and y gives

$$x = \sqrt{y + 2}$$
$$x^2 = y + 2$$

Step 2: Make y the subject.

$$y = x^2 - 2$$
$$f^{-1}(x) = x^2 - 2$$

Step 3: Use the relationship between f and f^{-1} to state the domain and range.

Since $\sqrt{x + 2} \geq 0$, the range of f is $f(x) \geq 0$. This means that the domain of f^{-1} is $x \geq 0$.

It is given that the domain of f is $x \geq -2$. This means that the range of f^{-1} is $f^{-1}(x) \geq -2$.

Step 4: Use the inverse function as the reverse process.

b $f(x) = \frac{1}{2}$

$$\Rightarrow \quad x = f^{-1}\left(\tfrac{1}{2}\right)$$
$$= \left(\tfrac{1}{2}\right)^2 - 2 = -\tfrac{7}{4}$$

So $\quad x = -\frac{7}{4}$.

Tip:
Square both sides.

Tip:
Make it clear for which function you are stating the domain or range.

Tip:
Often candidates set up the equation $f(x) = \frac{1}{2} \Rightarrow \sqrt{x+2} = \frac{1}{2}$ and solve it. That is an acceptable alternative method, but it may take longer.

Step 5: Sketch $y = f(x)$ for the domain, and reflect it in $y = x$.

c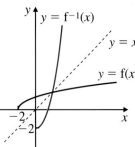

Step 6: Set up an equation in x and solve.

d From the graph, $f(x) = f^{-1}(x)$ at the point of intersection with the line $y = x$, i.e. when $f(x) = x$ and $f^{-1}(x) = x$.

Using $\quad f^{-1}(x) = x$ gives

$$x^2 - 2 = x$$
$$x^2 - x - 2 = 0$$
$$(x - 2)(x + 1) = 0$$
$$x = 2 \text{ or } -1$$

But $x \geqslant 0$, so $x = 2$.

In the above example, the equation $f(x) = f^{-1}(x)$ is solved by finding any points at which $f(x) = x$ or $f^{-1}(x) = x$, i.e. when the element is mapped back onto itself.

> **Tip:**
> It is much easier to solve $f(x) = x$ or $f^{-1}(x) = x$ than $f(x) = f^{-1}(x)$, for which you would have to solve $\sqrt{x + 2} = x^2 - 2$.
> Here $f^{-1}(x) = x$ is the easiest method as there are no $\sqrt{}$ signs to deal with.

> **Note:**
> Reject the negative value, since, from the sketch, the intersection occurs when $x \geqslant 0$.

Self-inverse

When $f(x) = f^{-1}(x)$ for *all* elements of the domain, the function is **self-inverse**, and f^{-1} is the same as f.

For example, consider the function f, defined by $f(x) = \dfrac{1}{x}, x \neq 0$.

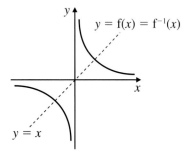

> **Note:**
> The reflection of $y = \dfrac{1}{x}$ in the line $y = x$ gives the same curve.

This is self-inverse because $f^{-1}(x) = \dfrac{1}{x}, x \neq 0$. So $f(x) = f^{-1}(x)$.

> **Note:**
> When a function is self-inverse, $ff^{-1}(x) = f^{-1}f(x) = x$.
> Also $ff(x) = x$.

SKILLS CHECK **1A: Functions**

1 State whether each of the following mappings represents a function, for real values of x. If it is a function, state whether it is one–one or many–one.

   a $y = 2(5 - x)$ **b** $y^2 = 1 - x^2$ **c** $y = -\dfrac{1}{x}, x \neq 0$ **d** $y = -x^2$

2 Find the range of each of the functions f defined by:

   a $f(x) = \dfrac{x + 5}{2}, x = \{0, 1, 2, 3\}$ **b** $f(x) = \dfrac{1}{5 - x}, x = \{1, 2, 3, 4\}$

   c $f(x) = x^3, x \geqslant 0$ **d** $f(x) = x^2 - 2, x \in \mathbb{R}$

3 For each of the functions defined below, **i** sketch the function, **ii** state its range:

    a $f(x) = 4x - 3$

     b $g(x) = \sin x°, 0 \leqslant x \leqslant 360$

    c $f(x) = \dfrac{1}{x^2}, x \in \mathbb{R}, x \neq 0$

    d $h(x) = x^2 - 6x$

 4 The function f is defined by $f(x) = \dfrac{6}{x} + 2x$ for the domain $1 \leqslant x \leqslant 3$. Find the range of f.

5 Functions f, g and h are defined for real values of x as follows:

$$f(x) = 1 - 2x \qquad g(x) = x^2 + 3 \qquad h(x) = \frac{x + 5}{2}$$

Find the value of

    a $fg(2)$ **b** $hf(-1)$ **c** $gg(0)$

    d $fh(-11)$ **e** $gf(\tfrac{1}{4})$ **f** $hgf(1.5)$

6 Functions f, g and h are defined for real values of x as follows:

$$f(x) = 2^x \qquad g(x) = 3x + 2 \qquad h(x) = \frac{1}{x}, x \neq 0$$

Find the composite functions

    a fg   **b** hh   **c** gh   **d** hg

7 Functions f, g and h are defined for real values of x as follows:

$$f(x) = 2x + 9 \qquad g(x) = \log_{10} x, x > 0 \qquad h(x) = 1 - x^2$$

Solve these equations:

    a $ff(x) = 9$ **b** $gf(x) = 0$ **c** $fh(x) = -5$ **d** $hf(x) = -8$

8 Which of the following functions has an inverse?

    a   **b**

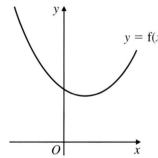 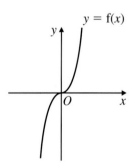

9 For each of the following functions, f:

    i find the inverse function, f^{-1}, stating its domain and range,

    ii on the same set of axes, sketch $y = f(x)$ and $y = f^{-1}(x)$.

    a $f(x) = 2x + 5, x \in \mathbb{R}$ **b** $f(x) = \dfrac{3 - x}{4}, x \in \mathbb{R}$

    c $f(x) = x^2, x \geqslant 0$ **d** $f(x) = \sqrt{x - 3}, 3 \leqslant x \leqslant 12$

10 Functions f, g and h are defined for all real values of x as follows:

$$f(x) = 5x - 4 \qquad g(x) = 1 - 2x \qquad h(x) = x^2$$

Solve these equations:

    a $gf(x) = g^{-1}(x)$ **b** $h(x) = g^{-1}(x)$ **c** $hg(x) = h(x)$

11 A function f is defined by $f(x) = 3 - \dfrac{2}{x}$, $x \neq 0$.

    a Find f^{-1} and state the value of x for which f^{-1} is undefined.

    b Find the values of x for which $f(x) = f^{-1}(x)$.

12 A function f is defined by $f(x) = \dfrac{1}{5 - 4x}$, $x \neq \frac{5}{4}$.

    a Find the values of x which map onto themselves under the function f.

    b Find an expression for f^{-1}.

Another function g is defined for real values of x by $g(x) = x^2 - 3$.

    c Evaluate $gf(1)$.

SKILLS CHECK **1A EXTRA is on the CD**

1.4 The modulus function

The modulus function.

The **modulus** (or absolute value or magnitude) of a number is its positive numerical value.

The modulus of x, $|x|$, is defined as

$$|x| = x, \qquad x \geq 0$$
$$|x| = -x, \qquad x < 0$$

For example,

$$|4| = |-4| = 4$$

Note:
The modulus is denoted by vertical lines as shown.

Tip:
Your calculator may have a modulus facility. It is often called 'abs'.

The graph of a modulus function

Consider the function f.

To draw the graph of $y = |f(x)|$, you need to consider the sign of $f(x)$.

When $f(x) \geq 0$, $|f(x)| = f(x)$, so the graph of $y = |f(x)|$ is the same as the graph of $y = f(x)$.

When $f(x) < 0$, $|f(x)| = -f(x)$, so the graph of $y = |f(x)|$ is a reflection in the x-axis of the graph of $y = f(x)$.

So to draw $y = |f(x)|$ from the graph of $y = f(x)$:

- draw any part of $y = f(x)$ that is on or above the x-axis
- replace any part of $y = f(x)$ that is below the x-axis with its reflection in the x-axis.

Tip:
When $f(x) > 0$, the graph of $y = f(x)$ is above the x-axis.

Tip:
When $f(x) < 0$, the graph of $y = f(x)$ is below the x-axis.

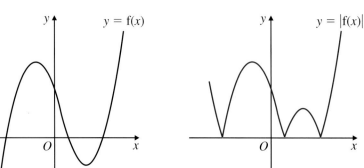

Graphical solution of equations and inequalities

Equations and inequalities involving modulus functions can be solved graphically as follows:

Example 1.7 **a** Sketch the graph of $y = 3x - 2$.

b The graph of $y = |3x - 2|$ consists of two parts. Draw a sketch to show the graph of $y = |3x - 2|$, labelling each part with its equation.

The line $y = 7$ intersects $y = |3x - 2|$ at P and Q.

c i Find the x-coordinates of P and Q.

ii Solve the inequality $|3x - 2| > 7$.

Step 1: Draw a sketch of $y = 3x - 2$.

a

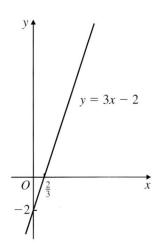

> **Tip:**
> When $x = 0$, $y = -2$ and when $y = 0$, $x = \frac{2}{3}$.

Step 2: Draw a second sketch, replacing the part of the line below the x-axis with its reflection in the x-axis and label the two parts separately.

b On the graph of $y = |3x - 2|$:

- when $3x - 2 \geqslant 0$, i.e. $x \geqslant \frac{2}{3}$, the equation of the line is $y = 3x - 2$

- when $3x - 2 < 0$, i.e. $x < \frac{2}{3}$, the equation of the line is $y = -(3x - 2)$.

> **Recall:**
> The reflection in the x-axis of $y = \mathrm{f}(x)$ is $y = -\mathrm{f}(x)$.

> **Note:**
> $y = -(3x - 2)$ could be written $y = 2 - 3x$.

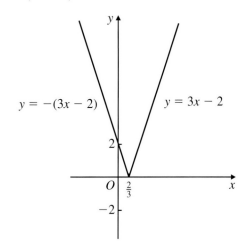

Step 3: Sketch the given line on the modulus graph, labelling the points *P* and *Q*. **c**

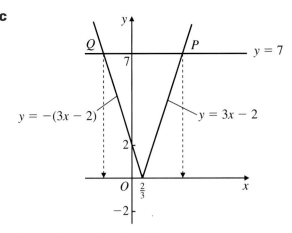

Step 4: Set up two equations and solve to find the *x*-coordinates of *P* and *Q*.

i At *P* $\quad 3x - 2 = 7$

$$3x = 9$$
$$x = 3$$

At *Q* $\; -(3x - 2) = 7$

$$3x - 2 = -7$$
$$3x = -5$$
$$x = -\tfrac{5}{3}$$

So the *x*-coordinates of *P* and *Q* are 3 and $-\tfrac{5}{3}$.

Step 5: Use the graphs to solve the inequality.

ii From the sketch:

$|3x - 2| > 7$ when $x < -\tfrac{5}{3}$ or $x > 3$.

Note:

If $|f(x)| = c$ then $f(x) = c$ or $-f(x) = c$.

Tip:

Find where the graph of $y = |3x - 2|$ is above the graph of $y = 7$.

Example 1.8 The diagram shows a sketch of $y = \sin x$, $0° \leqslant x \leqslant 360°$.

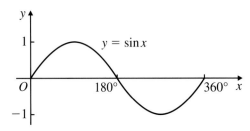

Recall:

Trigonometric graphs (C2 Section 3.5).

For $0° \leqslant x \leqslant 360°$:

a draw a sketch showing $y = |\sin x|$ and $y = 0.5$

b solve the equation $|\sin x| = 0.5$.

Step 1: Replace the given curve below the *x*-axis with its reflection in the *x*-axis.

a

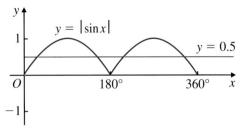

Step 2: Draw the given line on the same set of axes.

Tip:

Notice that two points of intersection are with the original part of the curve, i.e. $y = \sin x$, and the other two are with the reflected part of the curve, i.e. $y = -\sin x$.

Step 3: Solve the appropriate trig equations.

b $\quad |\sin x| = 0.5$

$$\Rightarrow \quad \sin x = 0.5$$
$$x = 30°, 150°$$

or $\; -\sin x = 0.5$

$$\sin x = -0.5$$
$$x = 210°, 330°$$

So $x = 30°, 150°, 210°, 330°$.

Recall:

Find the principal value from the calculator and then find other solutions in range (C2 Section 3.7).

Example 1.9 **a** On the same set of axes sketch the graphs of $y = |2x - 6|$ and
$y = |\frac{1}{2}x|$.

b Solve the equation $|2x - 6| = |\frac{1}{2}x|$.

Step 1: Draw a sketch of **a**
$y = |2x - 6|$.

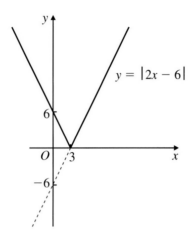

Tip:
Draw the graph of $y = 2x - 6$, but reflect the part below the x-axis in the x-axis.

Tip:
If you draw the line below the x-axis for guidance, make sure that it's clearly not part of your final answer.

Step 2: On the same diagram sketch $y = |\frac{1}{2}x|$.

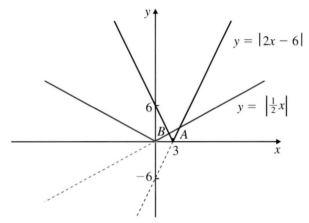

Tip:
The solutions for part **b** are the x-coordinates of the points of intersection, A and B.

Step 3: Set up two equations and solve.

b Intersection A is on the original part of both graphs, so

$$2x - 6 = \tfrac{1}{2}x$$
$$4x - 12 = x$$
$$3x = 12$$
$$x = 4$$

Intersection B is on the reflected part of $y = 2x - 6$,
i.e. on $y = -(2x - 6)$, and on the original part of $y = \frac{1}{2}x$, so

$$-(2x - 6) = \tfrac{1}{2}x$$
$$6 - 2x = \tfrac{1}{2}x$$
$$12 - 4x = x$$
$$5x = 12$$
$$x = 2.4$$

So $x = 4$ or $x = 2.4$

Tip:
Use your sketch to determine which branch of the modulus to use when solving the equation.

Tip:
Take care with negatives.

1.5 Combinations of transformations

Combinations of the transformations on the graph of $y = f(x)$ as represented by $y = af(x)$, $y = f(x) + a$, $y = f(x + a)$, $y = f(ax)$.

In *Core 2* you learnt that transformations applied to $y = f(x)$ have the following effects:

Recall:
Transformations (C2 Section 1.2).

13

Translations

$y = f(x) + a$ represents a translation by a units in the y-direction.

$y = f(x + a)$ represents a translation by $-a$ units in the x-direction.

Tip:
The vector form of a translation by a units in the y-direction is $\begin{bmatrix} 0 \\ a \end{bmatrix}$.

Stretches

$y = af(x)$ represents a stretch by scale factor a in the y-direction.

$y = f(ax)$ represents a stretch by scale factor $\dfrac{1}{a}$ in the x-direction.

Tip:
The vector form of a translation by $-a$ units in the x-direction is $\begin{bmatrix} -a \\ 0 \end{bmatrix}$.

Reflections

$y = -f(x)$ represents a reflection in the x-axis.

$y = f(-x)$ represents a reflection in the y-axis.

Combinations of transformations

In *Core* 3 you need to be able to apply **combinations** of these transformations. In some cases, the order in which the transformations are applied is important.

Tip:
You may be asked to apply transformations to the curves studied in C1 and C2 as well as curves introduced in C3.

Here are some examples:

$y = 2f(5x)$ is a stretch of $y = f(x)$ by scale factor $\frac{1}{5}$ in the x-direction and a stretch by scale factor 2 in the y-direction. If, for example, $f(x) = \sin x$, then $y = 2 \sin 5x$.

Note:
$y = f(x)$ is mapped to $y = af(kx)$.

$y = 2 - f(x)$ is a reflection of $y = f(x)$ in the x-axis and a translation by 2 units in the y-direction. The vector of the translation is $\begin{bmatrix} 0 \\ 2 \end{bmatrix}$.

If, for example, $f(x) = x^3$ then $y = 2 - x^3$.

Note:
$y = f(x)$ is mapped to $y = b - f(x)$.

$y = f(x - 4) + 3$ is a translation of $y = f(x)$ by 4 units in the x-direction and by 3 units in the y-direction. The vector of this translation is $\begin{bmatrix} 4 \\ 3 \end{bmatrix}$.

If, for example, $f(x) = x^2$ then $y = (x - 4)^2 + 3$.

Note:
$y = f(x)$ is mapped to $y = f(x - a) + b$.
You used this in C1 to sketch quadratic curves.
(C1 Section 1.15).

$y = 3f(x + 1)$ is a stretch of $y = f(x)$ by scale factor 3 in the y-direction and a translation by -1 unit in the x-direction.

If, for example, $f(x) = x^2$ then $y = 3(x + 1)^2$.

Note:
$y = f(x)$ is mapped to $y = af(x + b)$.

$y = 2f(x) + 5$ is a stretch by scale factor 2 in the y-direction followed by a translation by 5 units in the y-direction.

If, for example, $f(x) = x^2$, then $y = 2x^2 + 5$.

Note:
$y = f(x)$ is mapped to $y = af(x) + b$.

Note:
In this example the order is important. If the translation is carried out first, then $y = 2(x^2 + 5)$.

Sketching curves

Combinations of transformations can be used to sketch curves.

Example 1.10 The diagram shows a sketch of $y = f(x)$.

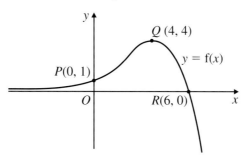

In each of the following, sketch the curve, showing the coordinates of P_1, Q_1 and R_1, the images of P, Q and R:

a $y = 4 - f(x)$ **b** $y = \frac{1}{2} f(x + 3)$ **c** $y = \frac{1}{4} f(2x)$

Step 1: Apply the transformations to the given curve.

a This is a reflection of $y = f(x)$ in the x-axis, followed by a translation of 4 units in the y-direction.

Tip:
The broken curve is the reflection. The solid curve is the final transformation. Make this clear on your sketch.

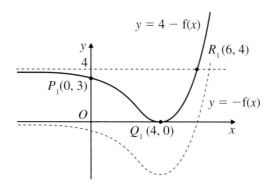

Note:
$y = 4$ is an asymptote of the curve $y = 4 - f(x)$.

Tip:
The reflection in the x-axis makes all the y-coordinates negative and then the vertical translation adds 4 to the y-coordinates. The x-values remain unchanged.

Step 2: Find the coordinates of the images of the given points.

The images of P, Q and R are $P_1(0, 3)$, $Q_1(4, 0)$ and $R_1(6, 4)$.

Step 1: Apply the transformations to the given curve.

b This is a translation of $y = f(x)$ by -3 units in the x-direction, and a stretch by scale factor $\frac{1}{2}$ in the y-direction.

Tip:
The horizontal translation takes 3 off the x-coordinates and the vertical stretch halves the y-coordinates.

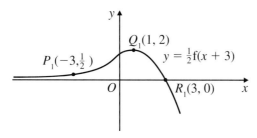

Step 2: Find the coordinates of the images of the given points.

The images of P, Q and R are $P_1(-3, \frac{1}{2})$, $Q_1(1, 2)$ and $R_1(3, 0)$.

Step 1: Apply the transformations to the given curve.

c This is a stretch by scale factor $\frac{1}{2}$ in the x-direction and a stretch by scale factor $\frac{1}{4}$ in the y-direction.

Tip:
The horizontal stretch halves the x-coordinates and the vertical stretch divides the y-coordinates by 4.

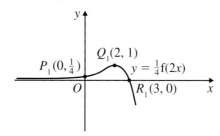

Step 2: Find the coordinates of the images of the given points.

The images of P, Q and R are $P_1(0, \frac{1}{4})$, $Q_1(2, 1)$ and $R_1(3, 0)$.

Example 1.11 The diagram shows a sketch of the graph $y = f(x)$, where $f(x) = \sin x$, $0 \leqslant x \leqslant 2\pi$.

Recall:
Trigonometric graphs (C2 Section 3.5).

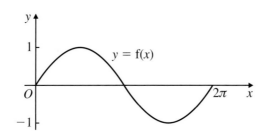

Given that $g(x) = 3 + \sin 2x$, $0 \leqslant x \leqslant 2\pi$,

a sketch the graph of $y = g(x)$

b state the range of g.

Step 1: Apply the transformations to the given curve.

a This is a stretch by scale factor $\frac{1}{2}$ in the x-direction, followed by a translation by 3 units in the y-direction.

Tip:
$y = \sin 2x$ is shown by the broken curve. If you show intermediate steps, make it clear which graph is which.

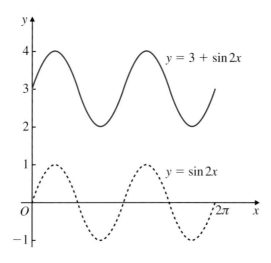

Tip:
The range of $\sin x$ is $-1 \leqslant y \leqslant 1$ so, following the translation by 3 units in the y-direction, this becomes $2 \leqslant y \leqslant 4$. Notice that the horizontal stretch doesn't affect the range.

Step 2: Use the sketch to find the range.

b From the graph, the range of g is $2 \leqslant g(x) \leqslant 4$.

You may also be asked to apply transformations to modulus functions.

For example, consider the graph of $y = 2|x| + 3$.

This is a stretch of $y = |x|$ by scale factor 2 in the y-direction to give $y = 2|x|$ followed by a translation of $y = 2|x|$ by 3 units in the y-direction.

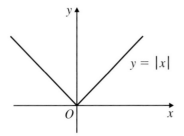

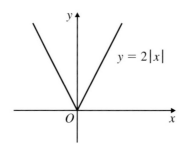

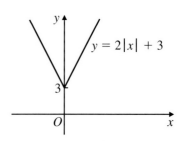

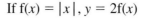
If $f(x) = |x|$, $y = 2f(x)$ If $g(x) = 2|x|$, $y = g(x) + 3$

1 For each of the following, sketch the graph, giving the coordinates of any points of intersection with the coordinate axes.

a $y = |2x - 3|$ **b** $y = |1 - 2x|$ **c** $y = |x| + 4$

2 For each of the functions f defined as follows

a $f(x) = x(x + 4)$ **b** $f(x) = \dfrac{1}{x}$ **c** $f(x) = \cos x, 0 \leqslant x \leqslant 2\pi$

on separate sets of axes, sketch the graph of

i $y = f(x)$ **ii** $y = |f(x)|$

showing the coordinates of any points at which the curve has a stationary point or meets the axes.

3 On separate sets of axes, sketch the following graphs, where a is a positive constant, giving the coordinates of any points of intersection with the coordinate axes.

a $y = |x - a|$ **b** $y = |2x + a|$ **c** $y = |x + a| + a$

4 For each pair of functions f and g, defined as follows:

a $f(x) = |3 - 2x|$ and $g(x) = x + 1$

b $f(x) = |3x - 5|$ and $g(x) = |2x + 1|$

 c $f(x) = |x|$ and $g(x) = 2|x - a|$, where a is a positive constant

  i   on the same set of axes, sketch the graphs of $y = f(x)$ and $y = g(x)$;

  ii solve the equation $f(x) = g(x)$;

  iii solve the inequality $f(x) > g(x)$.

5 Solve

a $|2x + 6| = 8$ **b** $|2x + 6| < 8$ **c** $|2x + 6| \geqslant 8$

6 By sketching appropriate graphs, or otherwise, solve the inequality

$$|2x + 3| > |3x - 6|.$$

7 For each of the following, describe how the graph of $y = f(x)$ can be mapped to the graph of $y = g(x)$ by applying two transformations:

a $f(x) = x$, $g(x) = \frac{1}{2}x - 3$

b $f(x) = 2^x$, $g(x) = 0.4(2^{-x})$

 c $f(x) = \dfrac{1}{x}$, $g(x) = \dfrac{5}{x - 2}$

8 A sketch of the graph of $y = f(x)$ is shown in the diagram.
The point P has coordinates $(-2, 0)$, Q has coordinates $(0, 3)$ and R has coordinates $(2, 0)$.

On separate sets of axes, sketch the graphs of the following, showing the coordinates of any points at which the curves have a stationary point.

a $y = f(x - 3) + 2$ **b** $y = \frac{1}{3}f(2x)$ **c** $y = -|f(x)|$

17

9 a The graph of $y = \tan x$ is subjected to a stretch in the y-direction by scale factor 3 followed by a translation by -2 units in the y-direction. Write down the equation of the resulting curve.

b The graph of $y = \sin x$ is subjected to a stretch in the x-direction by scale factor 3 followed by a reflection in the x-axis. Write down the equation of the resulting curve.

10 a Sketch the graph of $y = f(x)$, where $f(x) = \cos x$, $0 \leqslant x \leqslant 2\pi$.

Given that $g(x) = -\cos(x + \frac{1}{4}\pi)$, $0 \leqslant x \leqslant 2\pi$,

b sketch the graph of $y = g(x)$, showing the coordinates of any stationary points and intersections with the x-axis,

c state the range of g.

SKILLS CHECK **1B EXTRA is on the CD**

Examination practice 1:   Algebra and functions

1 The function g has domain $-1 \leqslant x \leqslant 2$ and is defined by $g(x) = x^2 + 5$.

a Find $g(-1)$ and $g(2)$.

b Sketch the graph of $y = g(x)$.

c Find the range of g.

d State, with a reason, whether the inverse function, g^{-1}, exists.

e Find $gg(x)$, giving your answer in the form $x^4 + px^2 + q$. [AQA Jan 2004]

2 The function f has domain $0 \leqslant x \leqslant 9$ and is defined by $f(x) = \sqrt{x} - 2$.

a i Find $f(0)$ and $f(9)$.

ii The graph of $y = \sqrt{x}$ for $x \geqslant 0$ is sketched below.

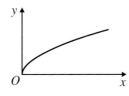

Hence sketch the graph of $y = f(x)$, stating clearly the values of the intercepts on the coordinate axes.

b Find the range of f.

c The inverse of f is f^{-1}.
i Find $f^{-1}(x)$.   **ii** State the domain of f^{-1}.   **iii** Sketch the graph of $y = f^{-1}(x)$.

[AQA Nov 2004]

 3 The function f is defined for all real values of x by $f(x) = x^2 + 6x + 11$.

a i Express $f(x)$ in the form $(x + p)^2 + q$, where p and q are integers.

ii State the value of x for which $f(x)$ is least.

iii Find the range of f.

b Solve the inequality $f(x) > 3$.

c State the geometrical transformation which maps the curve $y = f(x)$ onto:
i $y = f(x + 2)$;   **ii** $y = 4f(3x)$. [AQA May 2002]

4 a State which of the following graphs G_1, G_2 or G_3 does **not** represent a function. Give a reason for your answer.

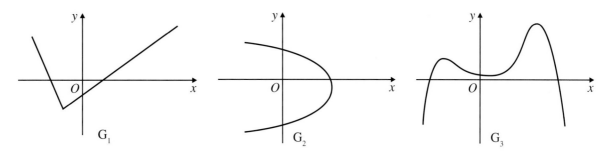

b The function f has domain $x \geq 2$ and is defined by $f(x) = \dfrac{1}{1-x} + 5$.

A sketch of $y = f(x)$ is shown below.

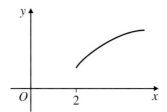

i Calculate $f(2)$ and $f(101)$.

ii Find the range of f.

iii The inverse of f is f^{-1}. Find $f^{-1}(x)$.

[AQA Nov 2002]

5 a The functions f and g are defined by:

$$f(x) = \sqrt{x} \qquad \text{for } x \geq 0;$$
$$g(x) = x - 1 \qquad \text{for all values of } x.$$

i Write down expressions for $fg(x)$ and $gf(x)$.

ii Verify that

$$x = 1 \Rightarrow fg(x) = gf(x).$$

b The diagram shows the graph of $y = h(x)$, where the function h is defined **for the domain $1 \leq x \leq 5$** by

$$h(x) = \sqrt{x - 1}$$

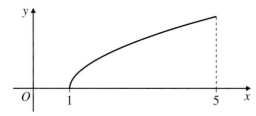

i Describe the transformation by which the graph of $y = \sqrt{x - 1}$ can be obtained from the graph of $y = \sqrt{x}$.

ii Write down the range of the function h.

iii Write down the domain and range of the inverse function h^{-1}.

iv Find an expression for $h^{-1}(x)$.

[AQA Jan 2004]

6 a Sketch the graph of $y = |8x|$.

b Sketch the graph of $y = \dfrac{1}{x^2}$, $x \neq 0$.

c i Verify that $x = \frac{1}{2}$ is a solution of the equation $\dfrac{1}{x^2} - |8x| = 0$.

ii The graphs of $y = \dfrac{1}{x^2}$ and $y = |8x|$ intersect at two points A and B. Find the coordinates of A and B. [AQA Nov 2004]

 7 The functions f and g are defined for real values of x as follows:

$$f(x) = a - x, \qquad g(x) = |2x + a|,$$

where a is a positive constant.

a Find gf($4a$).

b On the same set of axes, sketch the graphs of $y = f(x)$ and $y = g(x)$.

c Solve the equation $f(x) = g(x)$.

8 a Sketch the graph of $y = |2x - 4|$. Indicate the coordinates of the points where the graph meets the coordinate axes.

b i The line $y = x$ intersects the graph of $y = |2x - 4|$ at two points P and Q. Find the x-coordinates of the points P and Q.

ii Hence solve the inequality $|2x - 4| > x$.

c The graph of $y = |2x - 4| + k$ touches the line $y = x$ at only one point. Find the value of the constant k. [AQA Jan 2004]

9 The function f is defined for all real values of x by

$$f(x) = 3 - |2x - 1|$$

a i Sketch the graph of $y = f(x)$. Indicate the coordinates of the points where the graph crosses the coordinate axes.

ii Hence show that the equation $f(x) = 4$ has no real roots.

b State the range of f.

c By finding the values of x for which $f(x) = x$, solve the inequality

$$f(x) < x$$ [AQA Jan 2003]

 10 The diagram shows a sketch of the curve with equation $y = f(x)$. The curve crosses the x-axis at $A(-1, 0)$ and the y-axis at $B(0, 1)$.

Give the coordinates of A_1 and B_1, the images of A and B under the following transformations:

a $y = 3f(2x)$ **b** $y = 2 - f(x)$ **c** $y = f^{-1}(x)$

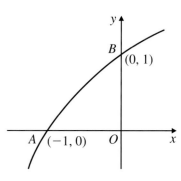

2 Trigonometry

2.1 Inverse trigonometric functions

Knowledge of $\sin^{-1} x$, $\cos^{-1} x$ and $\tan^{-1} x$. Understanding of their domains and graphs.

The three trigonometric functions of sine, cosine and tangent are many–one functions, since, for example,

$$\sin 30° = \sin 150° = \sin 390° = \cdots = 0.5$$

Since they are many–one functions, they do not have an inverse.

However, if we restrict the domain it is possible to define the **inverse trigonometric functions**: $\sin^{-1} x$, $\cos^{-1} x$ and $\tan^{-1} x$.

The graphs of the inverse functions are shown below. They have been drawn using the following properties of functions:

For a function f and its inverse function f^{-1}:

- the domain of f^{-1} is the range of f
- the range of f^{-1} is the domain of f
- the graph of $y = f^{-1}(x)$ is a reflection in the line $y = x$ of the graph of $y = f(x)$.

Recall:
Only one–one functions have an inverse (Section 1.3).

Note:
You can use degrees or radians.

Note:
The inverse trigonometric functions are also written $\arcsin x$, $\arccos x$ and $\arctan x$.

Note:
To show the reflection, the scales on both axes must be the same, so radians are used in the diagrams below.

$\sin^{-1} x$

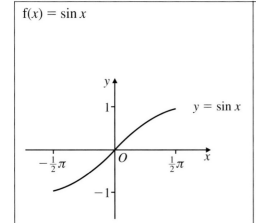

$f(x) = \sin x$

Domain $\quad -\tfrac{1}{2}\pi \leqslant x \leqslant \tfrac{1}{2}\pi$
Range $\quad -1 \leqslant \sin x \leqslant 1$

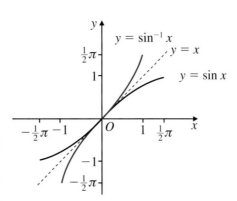

Reflect $y = \sin x$ in the line $y = x$:

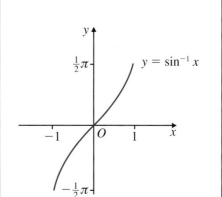

$f^{-1}(x) = \sin^{-1} x$

Domain $\quad -1 \leqslant x \leqslant 1$
Range $\quad -\tfrac{1}{2}\pi \leqslant \sin^{-1} x \leqslant \tfrac{1}{2}\pi$

Notice that if you turn your page through a quarter turn clockwise, and imagine the axis that is now horizontal as the x-axis, you can see a reflection in that axis of the sine curve.

cos⁻¹ x

$f(x) = \cos x$	Reflect $y = \cos x$ in the line $y = x$:	$f^{-1}(x) = \cos^{-1} x$
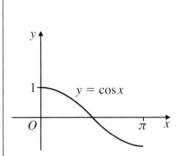		
Domain $\quad 0 \leqslant x \leqslant \pi$ Range $\quad -1 \leqslant \cos x \leqslant 1$		Domain $\quad -1 \leqslant x \leqslant 1$ Range $\quad 0 \leqslant \cos^{-1} x \leqslant \pi$

Notice that if you turn your page through a quarter turn clockwise, and imagine the axis that is now horizontal as the x-axis, you can see a reflection in that axis of the cosine curve.

tan⁻¹ x

$f(x) = \tan x$	Reflect $y = \tan x$ in the line $y = x$.	$f^{-1}(x) = \tan^{-1} x$
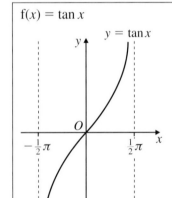	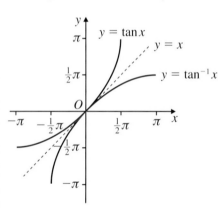	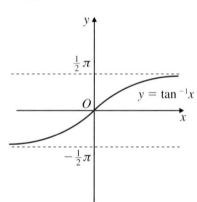
Domain $\quad -\frac{1}{2}\pi < x < \frac{1}{2}\pi$ Range $\quad \tan x \in \mathbb{R}$		Domain $\quad x \in \mathbb{R}$ Range $\quad -\frac{1}{2}\pi < \tan^{-1} x < \frac{1}{2}\pi$

The lines $y = -\frac{1}{2}\pi$ and $y = \frac{1}{2}\pi$ are asymptotes to the curve $y = \tan^{-1} x$.

Notice that if you turn your page through a quarter turn clockwise, and imagine the axis that is now horizontal as the x-axis, you can see a reflection in that axis of the tan curve.

Principal value (PV)

Remember that the value of $\sin^{-1} x$, $\cos^{-1} x$ or $\tan^{-1} x$ is an **angle**. It is the value given on the calculator by the inverse trig functions and is often referred to as the **principal value** (PV).

> **Recall:**
> The PV is used when solving trig equations. You can use degrees or radians (C2 Section 3.7).

Example 2.1 **a** Use a calculator to find, in degrees, the value of:

    i $\sin^{-1} 1$ **ii** $\cos^{-1} 0.5$

    iii $\tan^{-1}(-1)$ **iv** $\cos^{-1}(-0.3)$

b Use a calculator to find, in radians to two decimal places, the value of:

    i $\sin^{-1}(-0.6)$ **ii** $\cos^{-1} 0.3$ **iii** $\tan^{-1} 10$

Step 1: Use appropriate inverse trig functions on the calculator.

a **i** $\sin^{-1} 1 = 90°$ **ii** $\cos^{-1} 0.5 = 60°$

     iii $\tan^{-1}(-1) = -45°$ **iv** $\cos^{-1}(-0.3) = 107.45...°$

b **i** $\sin^{-1}(-0.6) = -0.6435...^c = -0.64^c$ (2 d.p.)

     ii $\cos^{-1} 0.3 = 1.2661...^c = 1.27^c$ (2 d.p.)

     iii $\tan^{-1} 10 = 1.4711...^c = 1.47^c$ (2 d.p.)

> **Tip:**
> For part **a**, set your calculator to degrees mode.

> **Tip:**
> For part **b**, set your calculator to radians mode.

2.2 Reciprocal trigonometric functions

Knowledge of secant, cosecant and cotangent functions to cosine, sine and tangent. Their relationship to sine, cosine and tangent.

The **reciprocal functions** of the three main trig functions of cosine (cos), sine (sin) and tangent (tan) are **secant** (sec), **cosecant** (cosec) and **cotangent** (cot). They are defined as follows:

$$\sec x = \frac{1}{\cos x} \qquad \operatorname{cosec} x = \frac{1}{\sin x} \qquad \cot x = \frac{1}{\tan x}$$

Since $\tan x = \dfrac{\sin x}{\cos x}$, we can also write $\cot x = \dfrac{\cos x}{\sin x}$.

> **Note:**
> Do not confuse the notation for the *reciprocal* function with the *inverse* function. For example, cosec x can be written $(\sin x)^{-1}$ whereas the inverse sine function is written $\sin^{-1} x$.

Graphs of reciprocal functions

You should learn the properties of the graphs of $y = \sec x$, $y = \operatorname{cosec} x$ and $y = \cot x$. It is useful to understand how they are obtained from the graphs of $y = \sin x$, $y = \cos x$ and $y = \tan x$ and these are also shown in the diagrams below.

> **Note:**
> A graphical calculator may be used in C3, so you can check the graphs.

y = sec x

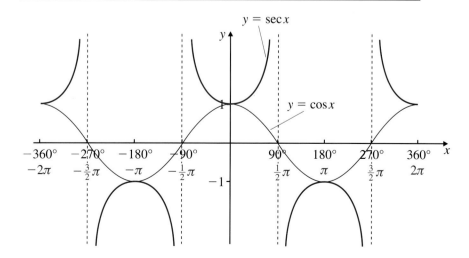

The value of sec x is always greater than or equal to 1 or less than or equal to -1.

There are vertical asymptotes through the points where $y = \cos x$ crosses the x-axis.

There are minimum points where $y = \cos x$ has maximum points.

There are maximum points where $y = \cos x$ has minimum points.

$y = \text{cosec } x$

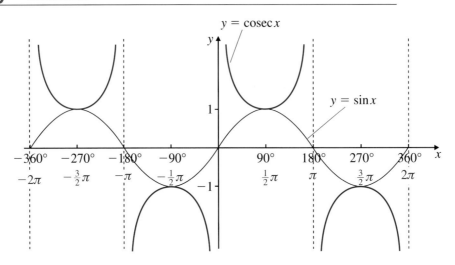

Whereas $\sin x$ takes values between -1 and 1, cosec x is always greater than or equal to 1 or less than or equal to -1.

There are vertical asymptotes through the points where $y = \sin x$ crosses the x-axis.

There are minimum points where $y = \sin x$ has maximum points.

There are maximum points where $y = \sin x$ has minimum points.

$y = \text{cot } x$

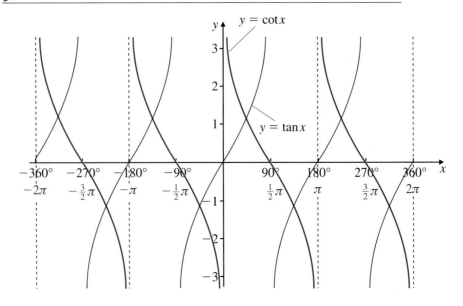

cot x can take all values.

The graph of $y = \cot x$ has vertical asymptotes through the points where $y = \tan x$ crosses the x-axis.

The graph of $y = \cot x$ crosses the x-axis where $y = \tan x$ has vertical asymptotes.

Applying transformations

You could be asked to apply combinations of transformations to the graphs of $y = \sec x$, $y = \operatorname{cosec} x$ and $y = \cot x$.

Recall:
Transformations (Section 1.5).

Example 2.2 $f(x) = 2 \sec x + 1$.

    a The graph of $y = f(x)$ can be obtained from the graph of $y = \sec x$ by applying a stretch followed by a translation.

        i State the scale factor and direction of the stretch.

        ii Describe the translation.

        iii State the coordinates of the image of the point $(0, 1)$ under this mapping.

    b Write down the range of f for $-90° < x < 90°$.

Step 1: Compare with $y = af(x) + b$ and describe the stretch and the translation.

Step 2: Apply the transformations in turn to the given point.

    a **i** The stretch is in the y-direction and the scale factor is 2.

        ii The translation is by 1 unit in the y-direction.

        iii Under the stretch, the image of $(0, 1)$ is $(0, 2)$.
Under the translation, the image of $(0, 2)$ is $(0, 3)$.
So the image of $(0, 1)$ is $(0, 3)$.

Step 3: Identify the range of f from the information in **a**.

    b The range of f for $-90° < x < 90°$ is $f(x) \geqslant 3$.

Note:
The vector of the translation is $\begin{bmatrix} 0 \\ 1 \end{bmatrix}$.

Tip:
For $-90° < x < 90°$, the minimum point of $y = \sec x$ is $(0, 1)$ and the range is $\sec x \geqslant 1$. What is the minimum point of $y = 2 \sec x + 1$?

2.3 Identities and equations involving the reciprocal functions

Knowledge and use of $1 + \tan^2 x = \sec^2 x$ and $1 + \cot^2 x = \operatorname{cosec}^2 x$.

In *Core 2* you used this identity:

$$\cos^2 \theta + \sin^2 \theta \equiv 1 \qquad ①$$

Manipulating this identity gives two further identities which must be learnt for *Core 3*:

$$1 + \tan^2 \theta \equiv \sec^2 \theta \qquad ②$$

$$\cot^2 \theta + 1 \equiv \operatorname{cosec}^2 \theta \qquad ③$$

Remember that these are identities, so they are true for all values of θ, in degrees or in radians.

You may be asked to use these identities to prove further identities or solve equations as in the following examples.

Recall:
Trig identities (C2 Section 3.6).

Tip:
For ②, divide each term of identity ① by $\cos^2 \theta$.

Tip:
For ③, divide each term of identity ① by $\sin^2 \theta$.

Example 2.3 Solve the equation

$$\sec^2 \theta = 5(\tan \theta - 1),$$

where $0° \leqslant \theta \leqslant 360°$, giving your answers in degrees, to the nearest degree.

Tip:
Set your calculator to degrees mode.

Step 1: Use an appropriate identity to form an equation in $\tan \theta$.

Step 2: Solve the equation in $\tan \theta$.

$$\sec^2 \theta = 5(\tan \theta - 1)$$
$$\Rightarrow \quad 1 + \tan^2 \theta = 5 \tan \theta - 5$$
$$\Rightarrow \quad \tan^2 \theta - 5 \tan \theta + 6 = 0$$
$$(\tan \theta - 3)(\tan \theta - 2) = 0$$
$$\Rightarrow \quad \tan \theta - 3 = 0$$
$$\tan \theta = 3$$
$$\theta = 71.5\ldots°,$$
$$251.5\ldots°$$

or
$$\tan \theta - 2 = 0$$
$$\tan \theta = 2$$
$$\theta = 63.4\ldots°,$$
$$243.4\ldots°$$

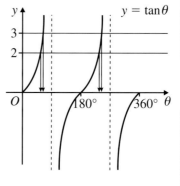

Tip:
The linear term is $\tan \theta$, so write $\sec^2 \theta$ in terms of $\tan \theta$.

Tip:
This is a quadratic equation in $\tan \theta$.

Tip:
Since the tan function repeats every 180°, the values of θ in range are PV and PV $+$ 180° (C2 Section 3.7).

So, to the nearest degree, $\theta = 63°, 72°, 243°, 252°$.

Example 2.4 Find, in radians to two decimal places, the values of x, where $-\pi \leqslant x \leqslant \pi$, such that
$$2 \cot^2 x = 3 \operatorname{cosec} x$$

Step 1: Use an appropriate identity to form an equation in $\operatorname{cosec} x$.

$$2 \cot^2 x = 3 \operatorname{cosec} x$$
$$2(\operatorname{cosec}^2 x - 1) = 3 \operatorname{cosec} x$$
$$2 \operatorname{cosec}^2 x - 2 = 3 \operatorname{cosec} x$$
$$2 \operatorname{cosec}^2 x - 3 \operatorname{cosec} x - 2 = 0$$

Step 2: Factorise and solve.

$$(2 \operatorname{cosec} x + 1)(\operatorname{cosec} x - 2) = 0$$
$$\Rightarrow \quad 2 \operatorname{cosec} x + 1 = 0$$
$$2 \operatorname{cosec} x = -1$$
$$\operatorname{cosec} x = -0.5 \text{ (no solutions)}$$

or
$$\operatorname{cosec} x - 2 = 0$$
$$\operatorname{cosec} x = 2$$
$$\sin x = 0.5$$
$$x = 0.523\ldots^c \text{ or } x = \pi - 0.523\ldots^c = 2.617\ldots^c$$

So, $x = 0.52^c$ (2 d.p.) or $x = 2.62^c$ (2 d.p.).

Tip:
The linear term is $\operatorname{cosec} x$, so aim to form an equation just in $\operatorname{cosec} x$ using the relationship $1 + \cot^2 x \equiv \operatorname{cosec}^2 x$.

Example 2.5 Prove the identity
$$(\sec x - \operatorname{cosec} x)(\sec x + \operatorname{cosec} x) \equiv (\tan x - \cot x)(\tan x + \cot x)$$

General steps for proving an identity:
Step 1: Start with one side of the identity and simplify/use appropriate identities to write it in a different format.
Step 2: Continue rewriting/simplifying until you get the expression on the other side of the identity.

$$\text{LHS} = (\sec x - \operatorname{cosec} x)(\sec x + \operatorname{cosec} x)$$
$$= \sec^2 x - \operatorname{cosec}^2 x$$
$$= (1 + \tan^2 x) - (1 + \cot^2 x)$$
$$= 1 + \tan^2 x - 1 - \cot^2 x$$
$$= \tan^2 x - \cot^2 x$$
$$= (\tan x - \cot x)(\tan x + \cot x)$$
$$= \text{RHS}$$

So $(\sec x - \operatorname{cosec} x)(\sec x + \operatorname{cosec} x) \equiv (\tan x - \cot x)(\tan x + \cot x)$.

Note:
In a proof, all working must be shown.

Recall:
Factorising the difference of two squares, where $(a - b)(a + b) = a^2 - b^2$ (C1 Section 1.4).

When proving an identity, you can start with either side. So you can start with the left-hand side and aim to get to the expression on the right-hand side, or vice versa. This gives a neat method of proof, but it is also acceptable to show that each side is equal to the same (third) expression. This is sometimes referred to as 'meeting in the middle'.

2A: Trigonometric functions, identities and equations

1 Use your calculator to find the value, to the nearest degree, of each of the following:

   a $\cos^{-1} 0.45$
   b $\sin^{-1}(-0.67)$
   c $\tan^{-1} 2.8$
   d $\cos^{-1}(-0.2)$

2 Use your calculator to find the value in radians, to two decimal places, of each of the following:

   a $\sin^{-1} 0.6$
   b $\tan^{-1}(-1.7)$
   c $\cos^{-1} 0.35$
   d $\cos^{-1}(-0.89)$

3 Solve the following equations, where $0° < x < 360°$. If your answer is not exact, give it to the nearest $0.1°$.

   a $\operatorname{cosec} x = 2.5$
   b $\cot x = -\sqrt{3}$
   c $\sec 2x = 1.5$
   d $\sec^2 x = 4$

4 Solve the following equations, where $0 \leqslant \theta \leqslant 2\pi$, giving your answers in radians, to three significant figures.

   a $\cot \frac{1}{2}\theta = 0.4$
   b $\operatorname{cosec} \theta = -1$
   c $\sec(\theta + \frac{1}{2}\pi) = 4$
   d $\operatorname{cosec}^2 \theta = 2$

 5 Prove these identities:

   a $\tan \theta + \cot \theta \equiv \sec \theta \operatorname{cosec} \theta$
      b $\sec \theta - \cos \theta \equiv \tan \theta \sin \theta$

 6 Solve $\tan^2 \theta = 1 + \sec \theta$ for $0° \leqslant \theta \leqslant 360°$.

7 Solve the equation

$$4 \operatorname{cosec} x - 5 = \cot^2 x$$

for $-360° \leqslant x \leqslant 360°$.

8 Solve the equation $\cot \theta = 2 \cos \theta$ for $0 < \theta \leqslant 2\pi$, giving your answers in radians, to two decimal places.

9 Find the values of x, where $-\pi \leqslant x \leqslant \pi$, such that $\operatorname{cosec}^2 x = 2 \cot x$, giving your answers in radians, to two decimal places.

10 $f(x) = -\cot(x + 90°)$.

   a Describe the geometrical transformations that map the graph of $y = \cot x$ onto the graph of $y = f(x)$.

   b On the same set of axes, sketch the graphs of $y = \cot x$ and $y = -\cot(x + 90°)$ for $-180° \leqslant x \leqslant 180°$, labelling each curve clearly.

   c Describe the relationship between $y = -\cot(x + 90°)$ and $y = \tan x$.

11 $f(x) = 2 \operatorname{cosec} x + 1$.

   a Describe the geometrical transformations that map the graph of $y = \operatorname{cosec} x$ onto the graph of $y = f(x)$.

   b State the coordinates of the image of the point $\left(\frac{1}{2}\pi, 1\right)$ under this mapping.

   c Write down the range of f for $0 < x < \pi$.

12 a Describe the geometrical transformations that map the graph of $y = \sec x°$ onto the graph of $y = 3 \sec 2x°$.

   b State the image of $(360, 1)$ under the mapping.

13 **a** State the equations of the asymptotes of the curve $y = \tan^{-1} x$.

 b Describe the geometrical transformations that map the graph of $y = \tan^{-1} x$ onto the graph of $y = 1 - \tan^{-1} x$.

 c State the equations of the asymptotes of the curve $y = 1 - \tan^{-1} x$.

SKILLS CHECK **2A EXTRA** is on the CD

Examination practice 2: Trigonometry

1 **a** Show that the equation

$$\tan^2 \theta + \sec \theta = 11$$

 can be written as

$$x^2 + x - 12 = 0$$

 where $x = \sec \theta$.

 b Hence solve the equation

$$\tan^2 \theta + \sec \theta = 11$$

 giving all the solutions to the nearest $0.1°$ in the interval $0° < \theta < 360°$. [AQA June 2003]

2 Solve $6 \cos x = 1 + \sec x$ for $0° < x < 360°$, giving your answers to one decimal place where necessary.

3 Solve the equation $3 \sec^2 x + \tan x - 5 = 0$ giving all the solutions in radians to two decimal places in the interval $0 < x < 2\pi$.

4 Solve the equation $\cot^2 x + 2 \operatorname{cosec} x - 7 = 0$ giving all the solutions to the nearest degree, where appropriate, in the interval $0° \leqslant x \leqslant 360°$.

5 Describe a sequence of geometrical transformations that maps the graph of $y = \tan x$ onto the graph of $y = 1 + \tan \frac{1}{2} x$.

6 **a** Solve the equation $3 \operatorname{cosec} x = 4 \sin x$ giving all the solutions in the interval $0° < x < 360°$.

 b Hence write down all the solutions in the interval $0° < x < 180°$ for which

$$3 \operatorname{cosec} 2x = 4 \sin 2x.$$

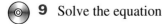 **7** Prove the identity

$$\operatorname{cosec}^2 \theta \cos^2 \theta \equiv \operatorname{cosec}^2 \theta - 1.$$

8 **a** Write down the equations of the asymptotes of the curve $y = \tan^{-1} x$.

 b The curve $y = \tan^{-1} x$ is subjected to a stretch in the y-direction by scale factor 2 to give the curve $y = f(x)$.

  i Write down the equation of the curve $y = f(x)$.

  ii Write down the equations of the asymptotes of the curve $y = f(x)$.

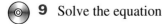 **9** Solve the equation

$$2 \operatorname{cosec}^2 x = 5(\cot x + 1)$$

 giving all the solutions in radians to two decimal places in the interval $0 < x < 2\pi$.

3 Exponentials and logarithms

3.1 The exponential function e^x

The function e^x and its graph.

In *Core 2* (Section 4.1) you met graphs of the form $y = a^x$, known as exponential curves.

Recall that

- when $x = 0$, $y = a^0 = 1$ so the graph passes through $(0, 1)$

- for $a > 1$, as $x \to \infty$, $y \to \infty$ and as $x \to -\infty$, $y \to 0$, so the x-axis is an asymptote.

Recall:
Exponent means index or power (C2 Section 1.1).

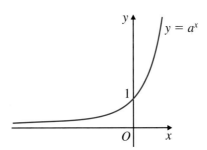

Remember that the graph of a^x becomes steeper than the polynomial graphs you have studied.

The function $f(x) = e^x$, where e is the irrational number 2.718..., is called **the** exponential function.

At the point $(0, 1)$, where $y = e^x$ cuts the y-axis, the curve has gradient 1.

Note:
e^x is sometimes written $\exp x$.

Note:
In fact, for all points on the curve $y = e^x$, $\dfrac{dy}{dx} = e^x$
(Section 4.1).

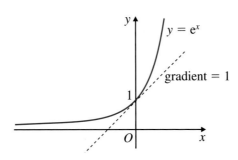

$e^x > 0$ for all values of x and, using the laws of indices, $e^0 = 1$.

Notice that as $x \to \infty$, $e^x \to \infty$ rapidly.

Recall:
$a^0 = 1$ (C2 Section 1.1).

Example 3.1 It is given that $f(x) = 2 + e^{-x}$.

a Sketch the graph of $y = f(x)$.

b State the equations of any asymptotes to the curve.

c State the domain and range of the function f.

Tip:
The word 'state' means you don't need to calculate these answers or show any working: you should just be able to write them down.

a Let $y = 2 + e^{-x}$.

The graph of $y = e^{-x}$ is a reflection in the y-axis of $y = e^x$.

The graph of $y = 2 + e^{-x}$ is a translation of $y = e^{-x}$ by 2 units in the y-direction.

When $x = 0$, $y = 2 + e^0 = 2 + 1 = 3$, so the curve cuts the y-axis at (0, 3).

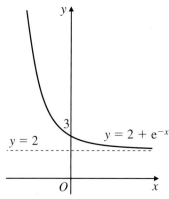

Recall:
$y = f(-x)$ is a reflection in the y-axis of $y = f(x)$; $y = f(x) + a$ translates the graph of $y = f(x)$ by a units in the y-direction (C1 Section 1.5).

Note:
e^{-x} is always greater than 0, so $2 + e^{-x}$ has to be greater than $2 + 0$, i.e. $y > 2$.

Tip:
As $x \to \infty$, $f(x) \to 2$, so don't let your graph turn upwards at the end.

b The x-axis is an asymptote to $y = e^{-x}$, so $y = 2$ is an asymptote to the translated curve.

c The domain of f is the set of real numbers and the range is the set of real numbers such that $y > 2$.

Recall:
For the function $y = f(x)$, the domain is the set of values that x can take. The range is the set of values that y can take. Both can be determined from the graph (Section 1.1).

3.2 The logarithmic function ln *x*

The function ln *x* and its graph. ln *x* as the inverse of e^x.

In *Core 2* you used the fact that $x = a^y \Leftrightarrow \log_a x = y$ to show that the inverse of an exponential function is a logarithmic function.

The inverse of the function e^x is $\ln x$, where **ln x** is the logarithm to the base e of x. It is known as the **natural log** of x.

So $x = e^y \Leftrightarrow \ln x = y$

Since $\ln x$ is the inverse of e^x, the graph of $y = \ln x$ is a reflection of the graph of $y = e^x$ in the line $y = x$.

The domain of $\ln x$ is the range of e^x, i.e. x is real and $x > 0$.
The range of $\ln x$ is the domain of e^x, i.e. y is the set of real numbers.

Note:
For base e, $x = e^y \Leftrightarrow \log_e x = y$

Recall:
Inverse functions (Section 1.3).

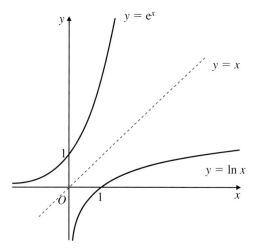

Using the laws of logs, $\ln 1 = 0$ and $\ln e = 1$.

Notice that as $x \to \infty$, $\ln x \to \infty$ slowly.

Recall:
$\log_a 1 = 0$, $\log_a a = 1$ (C2 Section 4.2).

Example 3.2 **a** Describe the two geometrical transformations by which the graph of $y = \frac{1}{2}\ln(x - 2) + 1$ can be obtained from the graph of $y = \ln x$.

b The line $y = a$ is an asymptote to the curve $y = \frac{1}{2}\ln(x - 2) + 1$. State the value of a.

c Sketch the graph of $y = \frac{1}{2}\ln(x - 2) + 1$, labelling the asymptote.

Step 1: Define the transformations.

a The graph of $y = \frac{1}{2}\ln(x - 2) + 1$ can be obtained from the graph of $y = \ln x$ by a stretch by scale factor $\frac{1}{2}$ in the y-direction followed by a translation by 2 units in the x-direction and 1 unit in the y-direction. The vector of the translation is $\begin{bmatrix} 2 \\ 1 \end{bmatrix}$.

Recall:
$y = af(x)$ is a stretch, scale factor a in y-direction.
$y = f(x - b) + c$ is a translation by $\begin{bmatrix} b \\ c \end{bmatrix}$ (Section 1.5).

Step 2: Consider any asymptotes.

b Since $y = \ln x$ has an asymptote at $x = 0$, $y = \frac{1}{2}\ln(x - 2) + 1$ must have an asymptote at $x = 2$.

Step 3: Draw in the asymptote.

c

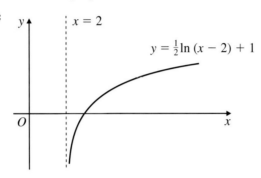

Step 4: Sketch the curve taking into account the stretch and translation.

Note:
The stretch must be carried out first. If the order of the transformations is reversed, the resulting graph is $y = \frac{1}{2}\ln(x - 2) + \frac{1}{2}$.

Example 3.3 Given that $f(x) = \ln(3 - x)$, $x < 3$, find $f^{-1}(x)$ and state the range of f^{-1}.

Tip:
Be careful!
$\ln(3 - x) \neq \ln 3 - \ln x$

Step 1: Let $y = f(x)$ and interchange x and y.

Let $y = \ln(3 - x)$

Now let $x = \ln(3 - y)$

$e^x = 3 - y$

Step 2: Make y the subject.

$y = 3 - e^x$

So $f^{-1}(x) = 3 - e^x$.

Step 3: State the range of the inverse function.

The range is $f^{-1}(x) < 3$.

Recall:
$\ln y = x \Leftrightarrow y = e^x$

Note:
You can interchange x and y at the end if you prefer.

Recall:
The range of f^{-1} is the domain of f (Section 1.3).

Solving equations involving e^x and $\ln x$

When solving equations involving e^x or $\ln x$, make use of the laws of indices and logarithms that you already know.

You will also need to use the fact that $\ln x$ is the inverse of e^x:

To solve an equation of the form $e^{ax + b} = p$, first take natural logs of both sides.

To solve an equation of the form $\ln(ax + b) = q$, first rewrite it as $ax + b = e^q$.

Also look out for cases when a substitution will transform the equation into a form that is easier to solve.

Recall:
Laws of indices (C2 Section 1.1).
Laws of logarithms (C2 Section 4.2).

Recall:
Natural logs are logs to the base e (Section 3).

Example 3.4 Find the *exact* solutions of the following equations:

a $e^{6x - 1} = 3$

b $e^x = 6e^{-x} + 5$

Step 1: Take natural logs of both sides.

a

$$e^{6x-1} = 3$$

$$\ln e^{6x-1} = \ln 3$$

Step 2: Use an appropriate log law.

$$(6x-1)\ln e = \ln 3$$

$$6x - 1 = \ln 3$$

Step 3: Rearrange the equation to find x.

$$6x = \ln 3 + 1$$

$$x = \frac{\ln 3 + 1}{6}$$

Recall:
$\log a^n = n \log a$ (C2 Section 4.2).

Recall:
$\ln e = 1$

Tip:
The word 'exact' here means leave your answers in terms of $\ln a$. Don't give your answer as a rounded decimal.

Step 1: Substitute for e^x.

b $e^x = 6e^{-x} + 5$

Let $y = e^x$, then $y = 6 \times \dfrac{1}{y} + 5$

$$= \frac{6}{y} + 5$$

Note:
If $y = e^x$, then $e^{-x} = \dfrac{1}{e^x} = \dfrac{1}{y}$.

Tip:
Don't forget to multiply **every** term by y.

Step 2: Multiply through by y.

$$y^2 = 6 + 5y$$

Step 3: Rearrange the equation and solve for y.

$$y^2 - 5y - 6 = 0$$

$$(y-6)(y+1) = 0$$

$$\Rightarrow \quad y = 6 \text{ or } y = -1$$

Tip:
If the expression does not factorise use the quadratic formula.

Step 4: Substitute back e^x for y.

$$e^x = 6 \text{ or } e^x = -1$$

Step 5: Solve for x, using $\ln x$ as the inverse of e^x.

Since $e^x > 0$ for all values of x, $e^x = -1$ has no solution.
So $e^x = 6 \Rightarrow x = \ln 6$.

Tip:
Don't forget to finish the question! Marks are often lost by candidates who forget to substitute back.

Example 3.5 Find the *exact* solutions of the following equations:

a $\ln(2y + 1)^2 = 6$

b $\ln(y + 1) - \ln y = 2$

Step 1: Use an appropriate log law.

a $\ln(2y + 1)^2 = 6$

$$2\ln(2y + 1) = 6$$

$$\ln(2y + 1) = \frac{6}{2} = 3$$

Recall:
$\log a^n = n \log a$

Tip:
$\ln y = x \Leftrightarrow y = e^x$

Step 2: Change the log to exponential form.

$$2y + 1 = e^3$$

Step 3: Rearrange to find y.

$$2y = e^3 - 1$$

$$y = \tfrac{1}{2}(e^3 - 1)$$

Note:
You don't have to simplify the log first; an alternative method would be
$(2y + 1)^2 = e^6$, so
$2y + 1 = \sqrt{e^6} = e^{\frac{6}{2}} = e^3$, as before.

Step 1: Simplify using the log laws.

b $\ln(y + 1) - \ln y = 2$

$$\ln \frac{y + 1}{y} = 2$$

Recall:
$\log x - \log y = \log \dfrac{x}{y}$
(C2 Section 4.2).

Step 2: Change the log to exponential form.

$$\frac{y + 1}{y} = e^2$$

Tip:
Be careful! $\ln \dfrac{y + 1}{y} \neq \dfrac{\ln (y + 1)}{\ln y}$.

Step 3: Eliminate the denominator.

$$y + 1 = e^2 y$$

Step 4: Make y the subject by rearranging and factorising.

$$e^2 y - y = 1$$

$$y(e^2 - 1) = 1$$

$$y = \frac{1}{e^2 - 1}$$

Tip:
From your calculator
$\dfrac{1}{e^2 - 1} = 0.1565\ldots$ Substitute back into the original equation to make sure you haven't made a slip.

1 Solve $e^{-\frac{x}{3}} = \frac{1}{2}$, giving your answer in the form $a \ln b$, where a and b are integers.

2 Using the substitution $y = e^x$, or otherwise, solve the equation $3e^{2x} = 2(e^x + 4)$.

3 Solve $e^{2x-5} = 1$.

4 Find the exact solution of $\ln(4x + 3) = 0.5$.

5 Use the substitution $y = \ln x$, or otherwise, to find the exact solutions of $(\ln x)^2 - 5\ln x + 6 = 0$.

6 Sketch each of the following on separate set of axes, stating the *exact* coordinates of any points of intersection with the x-axis and the equations of any asymptotes.

   a $y = |\ln x|, x > 0$

   b $y = \ln(x - 3) + 2, x > 3$

7 The function f is defined by

   a $f(x) = 3e^x$

   b $f(x) = \ln 2x, x > 0$

   For each function:

   i find the inverse function, f^{-1};

   ii sketch the graphs of $y = f(x)$ and $y = f^{-1}(x)$ on the same axes, showing the coordinates of any points of intersection with the axes;

   iii state the range of f and the domain and range of f^{-1}.

 8 The function f has domain $x > -1$ and is defined by

$$f(x) = \tfrac{1}{4}\ln(x + 1).$$

   a Find an expression for $f^{-1}(x)$.

   b State the domain and range of f^{-1}.

   c Find the exact value of x for which $f(x) = \frac{1}{2}$.

9 Given that $f(x) = e^{2x + 1}$,

   a find an expression for $f^{-1}(x)$;

   b sketch the graphs of $y = f(x)$ and $y = f^{-1}(x)$ on the same axes, stating the equations of any asymptotes;

   c state the range of f and the domain and range of f^{-1}.

10 a Describe a sequence of geometrical transformations by which the graph of $y = 2 - \ln(x + 1)$ can be obtained from the graph of $y = \ln x$.

   b Describe a sequence of geometrical transformations by which the graph of $y = 3e^{2x}$ can be obtained from the graph of $y = e^x$.

SKILLS CHECK **3A EXTRA is on the CD**

1   Solve each of the following equations, giving your answer exactly in terms of a natural logarithm:

   a $e^{-2x} = \dfrac{1}{16}$ **b** $e^{3x+6} = 8$

2   Solve the following equations, giving your answer in terms of e:

   a $\ln y - \ln(y - 1) = 1$ **b** $\ln(2x - 4) = 3$

3   **a** Sketch, on the same axes, the graphs with equations $y = 1 + e^{-x}$ and $y = 2\,|x + 4|$.

   b Write down the coordinates of any points where the graphs meet the axes.

The graphs intersect at the point where $x = p$.

   c Show that $x = p$ is a root of the equation $e^{-x} - 2x - 7 = 0$.

4   A function f is defined for real values of x by

      $f(x) = e^{x+1} - 2.$

   a Sketch the curve $y = f(x)$ and write down the exact coordinates of the points of intersection of the curve with the axes.

   b Find the inverse function, f^{-1}.

   c State the domain and range of f^{-1}.

5   **a** Sketch the graph of $y = k + \ln \dfrac{x}{2}$.

   b Find, in terms of k, the coordinates of the point of intersection with the x-axis.

   c Given that the curve crosses the x-axis at the point $\left(\dfrac{2}{e^2}, 0\right)$, show that $k = 2$.

6   Describe, in each of the following cases, a single transformation which maps the graph of $y = e^x$ onto the graph of the function given.

   a $y = e^{3x}$ **b** $y = e^{x-3}$ **c** $y = \ln x$ [AQA June 2001]

7   The diagram shows the graph of $y = 2e^{-x}$.

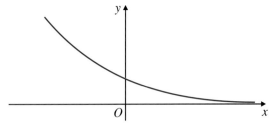

   a Describe a series of geometrical transformations by which the graph of $y = 2e^{-x}$ can be obtained from that of $y = e^x$.

   b The function f is defined **for the restricted domain $x \geqslant 0$** by

       $f(x) = 2e^{-x}.$

     i   State the range of the function f.

     ii State the domain and range of the inverse function f^{-1}.

     iii Find an expression for $f^{-1}(x)$.

     iv State, giving a reason, whether

        $x > \ln 2 \Rightarrow f(x) < 1$ [AQA Nov 2002]

8 The diagram shows the graphs of $y = x$ and $y = f(x)$.

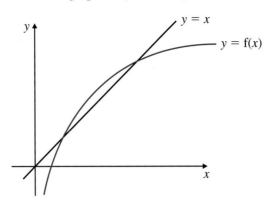

a i Describe the geometrical transformation by which the graph of $y = f^{-1}(x)$ can be obtained from the graph of $y = f(x)$.

ii Copy the above diagram and sketch on the same axes the graph of

$$y = f^{-1}(x).$$

b The function f is defined for $x > 0$ by

$$f(x) = 3 \ln x.$$

i Describe the geometrical transformation by which the graph of $y = f(x)$ can be obtained from the graph of $y = \ln x$.

ii Find an expression for $f^{-1}(x)$.

[AQA May 2002]

 9 a Show that $(x - 3)$ is a factor of $p(x) = x^3 - 2x^2 - 5x + 6$.

b Express $p(x)$ as a product of three linear factors.

c Solve the equation

$$e^{3y} - 2e^{2y} - 5e^y + 6 = 0$$

leaving your answers as exact values.

4 Differentiation

Differentiation of e^x, ln x, sin x, cos x, tan x and linear combinations of these.

In *Core 1* and *Core 2* you differentiated powers of x as follows:

$$y = ax^n \Rightarrow \frac{dy}{dx} = nax^{n-1}$$

You also used the fact that

$$y = f(x) \pm g(x) \Rightarrow \frac{dy}{dx} = f'(x) \pm g'(x)$$

This table shows the derivatives of some common functions needed in *Core 3*.

f(x)	f'(x)
e^x	e^x
ln x	$\dfrac{1}{x}$
sin x	cos x
cos x	$-\sin x$
tan x	$\sec^2 x$

Recall: Multiply by the power of x and decrease the power of x by 1 (C1 Section 3.2, C2 Section 5.1).

Recall: You can differentiate term by term.

Note: You should learn these results.

Note: If you forget the derivative of tan x it can be deduced from the derivative of tan kx given in the formulae booklet. It can also be found by using the quotient rule (Section 4.3).

Note: The trigonometric results are only true for x in radians.

Example 4.1 Differentiate with respect to x:

a $y = x^3 - 2\cos x$ **b** $A = 5e^x + 2\sqrt{x} + 6$

Step 1: Differentiate term by term.

a $\quad y = x^3 - 2\cos x$

$$\frac{dy}{dx} = 3x^2 - 2(-\sin x)$$

Step 2: Simplify.

$$= 3x^2 + 2\sin x$$

Step 3: Write powers of x in index form.

b $\quad A = 5e^x + 2\sqrt{x} + 6$

$$= 5e^x + 2x^{\frac{1}{2}} + 6$$

Step 4: Differentiate term by term.

$$\frac{dA}{dx} = 5e^x + 2(\tfrac{1}{2})x^{-\frac{1}{2}} + 0$$

$$= 5e^x + x^{-\frac{1}{2}}$$

Tip: It can help to wait until the end to simplify the signs, rather than try to do too many things at once.

Recall: The derivative of a constant is 0.

Example 4.2 It is given that $y = \ln(4x^3)$.

a Simplify $\ln(4x^3)$.   **b** Hence find $\dfrac{dy}{dx}$.

Step 1: Simplify the expression using log laws.

a $\quad y = \ln(4x^3)$

$$= \ln 4 + \ln x^3$$

$$= \ln 4 + 3\ln x$$

Step 2: Differentiate term by term.

b $\quad \dfrac{dy}{dx} = 0 + 3 \times \dfrac{1}{x} = \dfrac{3}{x}$

Tip: When functions involve logarithms (ln), always try to simplify, using the laws of logarithms, before differentiating.

Tip: $\ln(4x^3) \neq 3\ln(4x)$. Use $\log(xy) = \log x + \log y$ then $\log x^a = a\log x$ (C2 Section 4.2).

Example 4.3 A curve C, with equation $y = x^2 + 3e^x - 1$, crosses the y-axis at the point P. Find an equation of the normal to C at P, giving your answer in the form $ax + by = c$.

Step 1: Differentiate to find $\dfrac{dy}{dx}$.

$$y = x^2 + 3e^x - 1$$

$$\frac{dy}{dx} = 2x + 3e^x$$

Step 2: Substitute the x-value to get the gradient of the tangent at P.

When $x = 0$, $\dfrac{dy}{dx} = 2(0) + 3e^0 = 3$

Gradient of the tangent at P is 3

Tip:
Be careful when evaluating $3e^0$.
It is 3, not 0 or 1.

Step 3: Find the gradient of the normal at P.

$\Rightarrow$ gradient of normal at P is $-\frac{1}{3}$

Recall:
The tangent and normal are perpendicular, so the product of their gradients is -1
(C1 Section 2.2).

Step 4: Substitute x into the equation of the curve to get y.

When $x = 0$,

$$y = 0^2 + 3e^0 - 1 = 3 - 1 = 2$$

Step 5: Use an appropriate straight line equation.

Equation of the normal at $P(0, 2)$:

$$y - 2 = -\tfrac{1}{3}(x - 0)$$

$$y - 2 = -\tfrac{1}{3}x$$

$$3(y - 2) = -x$$

$$3y - 6 = -x$$

$$x + 3y = 6$$

Recall:
Equation of line
$y - y_1 = m(x - x_1)$
(C1 Section 2.1).

Note:
If you are not asked to give a specific form, leave the equation in a convenient form.

Example 4.4 The curve $y = 4 \sin x + 2x - 1$ is defined for $0 \leqslant x \leqslant \pi$ and has a stationary point at A.

a Find **i** $\dfrac{dy}{dx}$ **ii** $\dfrac{d^2y}{dx^2}$.

b **i** Find the x-coordinate of A.
    ii Determine whether A is a maximum or a minimum point.

Step 1: Differentiate to get $\dfrac{dy}{dx}$.

a **i** $y = 4 \sin x + 2x - 1$

$$\frac{dy}{dx} = 4 \cos x + 2$$

Step 2: Differentiate $\dfrac{dy}{dx}$ to get $\dfrac{d^2y}{dx^2}$.

ii $\dfrac{d^2y}{dx^2} = -4 \sin x$

Step 3: Set $\dfrac{dy}{dx} = 0$ and solve for x.

b **i**   At A, $\dfrac{dy}{dx} = 0$

$$\Rightarrow \quad 4 \cos x + 2 = 0$$

$$\cos x = -0.5$$

$$x = 2.094\ldots^c$$

Note:
The value of x could be given exactly. It is $x = \frac{2}{3}\pi$. This will not be required however in the examination.

Step 4: Substitute the x-value into $\dfrac{d^2y}{dx^2}$, check the sign and make your conclusion.

ii When $x = 2.094\ldots^c$,

$$\frac{d^2y}{dx^2} = -4 \sin (2.094\ldots^c) = -3.464\ldots < 0.$$

Since $\dfrac{d^2y}{dx^2} < 0$, A is a maximum point on the curve.

Recall:
$\dfrac{d^2y}{dx^2} < 0 \Rightarrow$ max point
$\dfrac{d^2y}{dx^2} > 0 \Rightarrow$ min point
(C1 Section 3.6,
C2 Section 5.3).

The **chain rule** is one of the most useful results in differentiation. It enables composite functions such as $(5x - 3)^7$, $4 \sin^2 x$, $\frac{1}{2} e^{3x+5}$ and $\ln \dfrac{1}{\sqrt{2x + 9}}$ to be differentiated.

Recall:
A composite function is formed by combining two or more functions (Section 1.2).

If y is a function of t and t is a function of x, then, by the chain rule,

$$\frac{dy}{dx} = \frac{dy}{dt} \times \frac{dt}{dx}$$

Note:
This is also referred to as 'differentiating a function of a function'.

An alternative version of the chain rule, using function notation, is

$$\frac{d}{dx} f(g(x)) = f'(g(x)) \times g'(x)$$

Example 4.5 Find $\dfrac{dy}{dx}$ when

a $y = (5x - 3)^7$ **b** $y = 4 \sin^2 x$

c $y = \frac{1}{2} e^{3x+5}$ **d** $y = \ln \dfrac{1}{\sqrt{2x + 9}}$

Step 1: Define t as a function of x; y is now a function of t.

Step 2: Differentiate t with respect to x, and y with respect to t.

Step 3: Rewrite t in terms of x.

Step 4: Apply the chain rule.

a $y = (5x - 3)^7$

Let $t = 5x - 3$, then $y = t^7$

So $\dfrac{dt}{dx} = 5$

and $\dfrac{dy}{dt} = 7t^6$

$\qquad = 7(5x - 3)^6$

$\dfrac{dy}{dx} = \dfrac{dy}{dt} \times \dfrac{dt}{dx}$

$\qquad = 7(5x - 3)^6 \times 5$

$\qquad = 35(5x - 3)^6$

Step 1: Define t as a function of x; y is now a function of t.

Step 2: Differentiate t with respect to x, and y with respect to t.

Step 3: Rewrite t in terms of x.

Step 4: Apply the chain rule.

b $y = 4 \sin^2 x = 4 (\sin x)^2$

Let $t = \sin x$, then $y = 4t^2$

So $\dfrac{dt}{dx} = \cos x$

and $\dfrac{dy}{dt} = 2 \times 4t$

$\qquad = 8 \sin x$

$\dfrac{dy}{dx} = \dfrac{dy}{dt} \times \dfrac{dt}{dx}$

$\qquad = 8 \sin x \times \cos x$

$\qquad = 8 \sin x \, \cos x$

Tip:
Rewrite the equation in this form. It will make it easier to see how to substitute.

Recall:
$\dfrac{d}{dx} (\sin x) = \cos x$

Step 1: Define *t* as a function of *x*; *y* is now a function of *t*.

c $y = \frac{1}{2}e^{3x+5}$

Let $t = 3x + 5$, then $y = \frac{1}{2}e^t$

Step 2: Differentiate *t* with respect to *x* and *y* with respect to *t*.

So $\dfrac{dt}{dx} = 3$

and $\dfrac{dy}{dt} = \frac{1}{2}e^t$

Step 3: Rewrite *t* in terms of *x*.

$= \frac{1}{2}e^{3x+5}$

Step 4: Apply the chain rule.

$\dfrac{dy}{dx} = \dfrac{dy}{dt} \times \dfrac{dt}{dx}$

$= \frac{1}{2}e^{3x+5} \times 3$

$= \frac{3}{2}e^{3x+5}$

Step 1: Simplify the expression using log laws.

d $y = \ln \dfrac{1}{\sqrt{2x+9}}$

$= \ln(2x+9)^{-\frac{1}{2}}$

$= -\frac{1}{2}\ln(2x+9)$

Step 2: Define *t* as a function of *x*; *y* is now a function of *t*.

Let $t = 2x + 9$, then $y = -\frac{1}{2}\ln t$

So $\dfrac{dt}{dx} = 2$

Step 3: Differentiate *t* with respect to *x*, and *y* with respect to *t*.

and $\dfrac{dy}{dt} = -\dfrac{1}{2} \times \dfrac{1}{t}$

Step 4: Rewrite *t* in terms of *x*.

$= -\dfrac{1}{2(2x+9)}$

Step 5: Apply the chain rule.

$\dfrac{dy}{dx} = \dfrac{dy}{dt} \times \dfrac{dt}{dx}$

$= -\dfrac{1}{2(2x+9)} \times 2$

$= -\dfrac{1}{2x+9}$

Example 4.6 The point $P(1, -1)$ lies on the curve with equation $y = \dfrac{1}{(2x-3)^5}$.

Find the gradient of the curve at P.

Step 1: Write the function of *x* in index form.

$y = \dfrac{1}{(2x-3)^5} = (2x-3)^{-5}$

Step 2: Define *t* as a function of *x*; *y* is now a function of *t*.

Let $t = 2x - 3$, then $y = t^{-5}$

Step 3: Differentiate *t* with respect to *x*, and *y* with respect to *t*.

So $\dfrac{dt}{dx} = 2$

and $\dfrac{dy}{dt} = -5t^{-6}$

Step 4: Rewrite *t* in terms of *x*.

$= -\dfrac{5}{(2x-3)^6}$

Step 5: Apply the chain rule.

$$\frac{dy}{dx} = \frac{dy}{dt} \times \frac{dt}{dx}$$

$$= -\frac{5}{(2x-3)^6} \times 2$$

$$= -\frac{10}{(2x-3)^6}$$

Step 6: Substitute the given value into $\frac{dy}{dx}$ and evaluate.

When $x = 1$, $\dfrac{dy}{dx} = -\dfrac{10}{(2 \times 1 - 3)^6} = -10$.

So the gradient of the curve at P is -10.

Some standard derivatives

It will save you time in the examination if you remember the following standard results, derived using the chain rule:

$f(x)$	$f'(x)$
e^{kx}	ke^{kx}
$\ln kx$	$\dfrac{1}{x}$
$\sin kx$	$k \cos kx$
$\cos kx$	$-k \sin kx$
$\tan kx$	$k \sec^2 kx$

Note:
Don't worry if you forget these; you can always work them out using the chain rule.

Note:
The derivative of $\tan kx$ is given in the formulae booklet. Check that you know where to find it.

Example 4.7 Given that $y = \sin^3 5x$, find $\dfrac{dy}{dx}$.

Step 1: Define t as a function of x; y is now a function of t.

$y = \sin^3 5x = (\sin 5x)^3$

Let $t = \sin 5x$, then $y = t^3$

Step 2: Differentiate t with respect to x and y with respect to t.

So $\dfrac{dt}{dx} = 5 \cos 5x$

and $\dfrac{dy}{dt} = 3t^2$

Step 3: Rewrite t in terms of x.

$= 3 \sin^2 5x$

Step 4: Apply the chain rule.

$\dfrac{dy}{dx} = \dfrac{dy}{dt} \times \dfrac{dt}{dx}$

$= 3 \sin^2 5x \times 5 \cos 5x$

$= 15 \cos 5x \sin^2 5x$

Tip:
If you had forgotten that
$\dfrac{d}{dx}(\sin kx) = k \cos kx$
you would need to apply the chain rule twice to find the derivative of $\sin^3 5x$.

Example 4.8 The curve with equation $y = 2e^{2x} - 4x$ has a stationary point at P.

a Determine $\dfrac{dy}{dx}$ and $\dfrac{d^2y}{dx^2}$ as functions of x.

b Find the coordinates of P.

c Find the value of $\dfrac{d^2y}{dx^2}$ at P and hence deduce the nature of the stationary point.

Step 1: Differentiate y with respect to x.	**a** $\quad y = 2e^{2x} - 4x$

$$\frac{dy}{dx} = 2 \times 2e^{2x} - 4 = 4e^{2x} - 4$$

Tip:
If you forget that $\frac{d}{dx}(e^{kx}) = ke^{kx}$ you could use the chain rule.

Step 2: Differentiate again. $\quad \dfrac{d^2y}{dx^2} = 4 \times 2e^{2x} = 8e^{2x}$

Step 3: Set $\dfrac{dy}{dx} = 0$ and solve for x.

b At P, $\quad \dfrac{dy}{dx} = 0$

Recall:
At a stationary point the gradient is zero (C1 Section 3.6).

$$\Rightarrow \quad 4e^{2x} - 4 = 0$$

$$4(e^{2x} - 1) = 0$$

$$e^{2x} - 1 = 0$$

$$e^{2x} = 1$$

Taking natural logs of both sides gives

$$\ln e^{2x} = \ln 1$$

Recall:
$\ln x$ is the inverse of e^x; $\ln 1 = 0$; $\ln e = 1$ (Section 3.2).

$$\Rightarrow \quad 2x \ln e = 0$$

$$2x = 0$$

$$x = 0$$

Step 4: Substitute the x-value into the equation of the curve to find y.

When $x = 0$, $y = 2e^{2 \times 0} - 4 \times 0 = 2 \times 1 - 0 = 2$, so P is the point $(0, 2)$.

Recall:
$e^0 = 1$ (Section 3.1).

Step 5: Substitute the x-coordinate of P into $\dfrac{d^2y}{dx^2}$.

c $\quad \dfrac{d^2y}{dx^2} = 8e^{2x}$

At P, $\quad x = 0 \Rightarrow \dfrac{d^2y}{dx^2} = 8e^{2 \times 0} = 8 \times 1 = 8$

Recall:
Nature of a stationary point (C1 Section 3.6, C2 Section 5.3).

Step 6: Determine the sign of $\dfrac{d^2y}{dx^2}$.

Since $\dfrac{d^2y}{dx^2} > 0$, P is a minimum point.

Example 4.9 It is given that $f(x) = \tan^2 x - 2\tan x$.

    a Find an expression in $\tan x$ for $f'(x)$.

    b Given that $\tan \frac{1}{3}\pi = \sqrt{3}$, find the exact value of $f'(\frac{1}{3}\pi)$.

    c Hence deduce whether f is increasing or decreasing when $x = \frac{1}{3}\pi$.

Step 1: Define t as a function of x; y is now a function of t.

a Let $y = \tan^2 x - 2\tan x = (\tan x)^2 - 2\tan x$

Let $t = \tan x$, then $y = t^2 - 2t$

Tip:
Writing $\tan^2 x$ as $(\tan x)^2$ will help you to see the substitution more clearly.

Step 2: Differentiate t with respect to x, and y with respect to t.

So $\quad \dfrac{dt}{dx} = \sec^2 x$

and $\quad \dfrac{dy}{dt} = 2t - 2$

Step 3: Rewrite t in terms of x.

$$= 2\tan x - 2$$

Step 4: Apply the chain rule.	$\dfrac{dy}{dx} = \dfrac{dy}{dt} \times \dfrac{dt}{dx}$	
	$= (2\tan x - 2) \times \sec^2 x$	

Recall:
Trig identities (Section 2.3).

Step 5: Write the derivative in terms of $\tan x$.

$$= 2(\tan x - 1)(1 + \tan^2 x)$$

So $\quad f'(x) = 2(\tan x - 1)(1 + \tan^2 x)$

Step 6: Substitute the given value of x into $f'(x)$.

b $\quad f'(\tfrac{1}{3}\pi) = 2(\tan \tfrac{1}{3}\pi - 1)(1 + \tan^2 \tfrac{1}{3}\pi)$

$$= 2(\sqrt{3} - 1)(1 + 3)$$

$$= 8(\sqrt{3} - 1)$$

Recall:
If $f'(a) > 0$ then f is an increasing function at $x = a$ and if $f'(a) < 0$ then f is a decreasing function at $x = a$ (C1 Section 3.5).

Step 7: Consider the sign of $f'(x)$ at the given x-value.

c Since $8(\sqrt{3} - 1) = 5.85... > 0$, $f'(\tfrac{1}{3}\pi) > 0$, so f is an increasing function when $x = \tfrac{1}{3}\pi$.

SKILLS CHECK **4A: Differentiation including the chain rule**

1 Find $\dfrac{dy}{dx}$ when

a $\quad y = 5e^x$

b $\quad y = \ln x^3$

c $\quad y = 4\cos x - \sin x$

d $\quad y = 2x^3 - e^x$

e $\quad y = \ln \dfrac{5}{x^2}$

f $\quad y = \dfrac{\tan x}{4}$

2 Find $\dfrac{dy}{dx}$ when

a $\quad y = 3\cos(x^2)$

b $\quad y = 3\cos^2 x$

c $\quad y = 3e^{x^2 - 5x}$

d $\quad y = 4\ln(5 - 2x)$

e $\quad y = \tan^2 3x$

f $\quad y = \dfrac{1}{e^x}$

3 It is given that $y = \ln \dfrac{5x}{\sqrt[3]{2x + 7}}$.

a Simplify y.

b Find $\dfrac{dy}{dx}$.

4 Differentiate with respect to x:

a $\quad y = (5x + 6)^4$

b $\quad z = \sqrt{2x - 1}$

c $\quad t = \dfrac{1}{x + 2}$

 5 Find an equation of the tangent to the curve $y = (\tfrac{1}{2}x^2 - 1)^3$ at the point where $x = -2$.

6 A curve has equation $y = 2\sin x + \cos x$. In the range $0 < x < \tfrac{1}{2}\pi$, it has a stationary point at A.

a Find the x-coordinate of A.

b Determine whether there is a maximum or minimum point at A.

7 The curve C has equation $y = x^2 - 8\ln x$, $x > 0$. Show that C has only one stationary point and find its coordinates.

8 A curve C has equation $y = 3e^x - 2\ln 5x$.

    a Find $\dfrac{dy}{dx}$.

    b Find an equation of the tangent to the curve at $x = 1$.

    c Find the exact y-coordinate of the point where this tangent crosses the y-axis. Give your answer in the form $a - \ln b$.

9 The curve C, with equation $y = \tan x - \sin x$, is defined for $-\frac{1}{2}\pi < x < \frac{1}{2}\pi$ and has a stationary point at P. Find the coordinates of P.

10 It is given that $f(x) = (e^x + 3)(e^{2x} - 5)$.

    a Expand the brackets.   **b** Find $f'(x)$.   **c** Evaluate $f''(0)$.

11 $f(x) = \dfrac{e^{3x} + 1}{e^{2x}}$

    a Simplify $f(x)$.

    b Find the gradient of the tangent to the curve $y = f(x)$ at the point where $x = 0$.

12 Differentiate with respect to t: **a** $x = \dfrac{3}{1-t}$ **b** $y = \ln(2t + 3)^4$

13 For each of the functions f, determine whether f is increasing or decreasing when $x = 2$.

    a $f(x) = e^x - x^2$ **b** $f(x) = \dfrac{1}{x} - 2\ln x$

You must show sufficient working to support your answer.

14 Given that $y = \sin 3x$, show that $\dfrac{d^2y}{dx^2} + 9y = 0$.

15 The curve with equation $y = e^{3x} - 6x$ has a stationary point at A.

    a Find the coordinates of A.

    b **i** Find $\dfrac{d^2y}{dx^2}$.   **ii** Hence determine the nature of the stationary point A.

SKILLS CHECK **4A EXTRA** is on the CD

4.3 The product and quotient rules

Differentiation using the product rule and the quotient rule.

The product rule

To differentiate a product of two functions (i.e. an expression formed by multiplying two functions together) use the **product rule**.

If $y = uv$, where u and v are functions of x, then

$$\frac{dy}{dx} = u\frac{dv}{dx} + v\frac{du}{dx}$$

An alternative version of the rule, using function notation, is

$$\frac{d}{dx}(f(x)g(x)) = f'(x)g(x) + f(x)g'(x)$$

Note:
In words: first × derivative of second + second × derivative of first.

Note:
If you use the alternative version of the rule your working will be in a different order, but the answer will be the same.

43

Example 4.10 Find $\dfrac{dy}{dx}$, simplifying your answer, when

a $y = (3x^2 - 2)(1 - 4x^3)$ **b** $y = 2x^4 \sin x$

a $y = (3x^2 - 2)(1 - 4x^3)$

Step 1: Define u and v in terms of x.

Let $u = 3x^2 - 2, v = 1 - 4x^3$

Step 2: Differentiate u and v with respect to x.

So $\dfrac{du}{dx} = 6x$ and $\dfrac{dv}{dx} = -12x^2$

Step 3: Apply the product rule.

$\dfrac{dy}{dx} = u\dfrac{dv}{dx} + v\dfrac{du}{dx}$

$= (3x^2 - 2) \times (-12x^2) + (1 - 4x^3) \times 6x$

Step 4: Simplify.

$= -12x^2(3x^2 - 2) + 6x(1 - 4x^3)$

$= 6x[-2x(3x^2 - 2) + (1 - 4x^3)]$

$= 6x(-6x^3 + 4x + 1 - 4x^3)$

$= 6x(1 + 4x - 10x^3)$

Note:
In part **a** the derivative could be found by multiplying out the brackets and then differentiating without using the product rule.

Tip:
It's good practice to tidy up expressions that result from the product rule, even if it is not specifically requested in the question.

b $y = 2x^4 \sin x$

Step 1: Define u and v in terms of x.

Let $u = 2x^4, v = \sin x$

Step 2: Differentiate u and v with respect to x.

So $\dfrac{du}{dx} = 8x^3$ and $\dfrac{dv}{dx} = \cos x$

Step 3: Apply the product rule.

$\dfrac{dy}{dx} = u\dfrac{dv}{dx} + v\dfrac{du}{dx}$

Stp 4: Simplify.

$= 2x^4 \times \cos x + \sin x \times 8x^3$

$= 2x^3(x \cos x + 4 \sin x)$

Tip:
Take out common factors.

Example 4.11 The curve C has equation $y = e^x \cos x$, $0 \leqslant x < \frac{1}{2}\pi$. Show that the curve has a stationary point when $\tan x = 1$.

$y = e^x \cos x$

Step 1: Define u and v in terms of x.

Let $u = e^x, v = \cos x$

Step 2: Differentiate u and v with respect to x.

So $\dfrac{du}{dx} = e^x$ and $\dfrac{dv}{dx} = -\sin x$

Step 3: Apply the product rule.

$\dfrac{dy}{dx} = u\dfrac{dv}{dx} + v\dfrac{du}{dx}$

$= e^x \times (-\sin x) + \cos x \times e^x$

Step 4: Simplify.

$= e^x (\cos x - \sin x)$

Step 5: Set $\dfrac{dy}{dx} = 0$.

Stationary points occur when $\dfrac{dy}{dx} = 0$

i.e. when $e^x (\cos x - \sin x) = 0$

$\Rightarrow$ $e^x = 0$ (no solution) or $\cos x - \sin x = 0$

$\Rightarrow$ $\cos x = \sin x$

$\Rightarrow$ $\tan x = 1$

Recall:
e^x can never be 0 (Section 3.1).

Recall:
$\dfrac{\sin x}{\cos x} = \tan x$ (C2 Section 3.6).

Note:
The question doesn't ask for the coordinates of the stationary point so you are not required to find them.

The quotient rule

To differentiate a quotient of two functions (i.e. an expression formed by dividing one function by another) use the **quotient rule**.

If $y = \dfrac{u}{v}$, where u and v are functions of x, then

$$\frac{dy}{dx} = \frac{v\dfrac{du}{dx} - u\dfrac{dv}{dx}}{v^2}$$

Note:

In words: bottom × derivative of top − top × derivative of bottom, all over bottom squared.

An alternative version of the rule, using function notation, is

$$\frac{d}{dx}\left(\frac{f(x)}{g(x)}\right) = \frac{f'(x)g(x) - f(x)g'(x)}{(g(x))^2}$$

Note:

If you use the alternative version of the rule your working will be in a different order, but the answer will be the same.

Example 4.12 Differentiate $\dfrac{e^{3x}}{x}$ with respect to x, simplifying your answer.

Step 1: Define u and v in terms of x.

Let $\quad y = \dfrac{e^{3x}}{x}$

Let $\quad u = e^{3x}, \, v = x$

Step 2: Differentiate u and v with respect to x.

So $\quad \dfrac{du}{dx} = 3e^{3x}$

and $\quad \dfrac{dv}{dx} = 1$

Recall:

$\dfrac{d}{dx}(e^{kx}) = ke^{kx}$ (Section 4.2).

Step 3: Apply the quotient rule.

$$\frac{dy}{dx} = \frac{v\dfrac{du}{dx} - u\dfrac{dv}{dx}}{v^2}$$

$$= \frac{x \times 3e^{3x} - e^{3x} \times 1}{x^2}$$

Step 4: Simplify.

$$= \frac{e^{3x}(3x - 1)}{x^2}$$

Note:

You could differentiate the expression as a product of e^{3x} and x^{-1} to give

$\dfrac{dy}{dx} = e^{3x}(-x^{-2}) + x^{-1}(3e^{3x})$,

which then simplifies to the same result.

Example 4.13 The curve $y = \dfrac{\ln x}{x}$ has a maximum point at P.

Find the coordinates of P, giving your answer in terms of e.

Step 1: Define u and v in terms of x.

$$y = \frac{\ln x}{x}$$

Let $\quad u = \ln x, \, v = x.$

Step 2: Differentiate u and v with respect to x.

So $\quad \dfrac{du}{dx} = \dfrac{1}{x}$

and $\quad \dfrac{dv}{dx} = 1$

Step 3: Apply the quotient rule.

$$\frac{dy}{dx} = \frac{v\dfrac{du}{dx} - u\dfrac{dv}{dx}}{v^2}$$

$$= \frac{x \times \dfrac{1}{x} - \ln x \times 1}{x^2}$$

Step 4: Simplify.

$$= \frac{1 - \ln x}{x^2}$$

Step 5: Set $\dfrac{dy}{dx} = 0$. At P, $\dfrac{dy}{dx} = 0$

$$\Rightarrow \quad \frac{1 - \ln x}{x^2} = 0$$

$$\Rightarrow \quad 1 - \ln x = 0$$

$$\ln x = 1$$

$$x = e^1 = e$$

Tip:
The numerator must be zero.

Recall:
$\ln x = a \Leftrightarrow x = e^a$ (Section 3.2).

Step 6: Substitute into the equation of the curve to find the y-coordinate.

When $x = e$,

$$y = \frac{\ln e}{e} = \frac{1}{e}$$

The coordinates of P are $\left(e, \dfrac{1}{e} \right)$.

Recall:
$\ln e = 1$ (Section 3.2).

Derivatives of the reciprocal trigonometric functions

The derivatives of the following trigonometric functions can be derived using the quotient rule.

$f(x)$	$f'(x)$
$\sec x$	$\sec x \tan x$
$\cot x$	$-\text{cosec}^2 x$
$\text{cosec}\, x$	$-\text{cosec}\, x \cot x$

Note:
These are given in the formulae booklet, so can be quoted unless you are asked to derive them.

Tip:
The derivatives of trigonometric functions starting with 'co' have a negative sign.

Example 4.14 Use the derivatives of $\sin x$ and $\cos x$ to prove that the derivative of $\cot x$ is $-\text{cosec}^2 x$.

Note:
The question asks for a proof, so you need to show all the steps in your working.

Step 1: Write $\cot x$ in terms of $\sin x$ and $\cos x$.

$$y = \cot x = \frac{1}{\tan x} = \frac{\cos x}{\sin x}$$

Step 2: Define u and v in terms of x.

Let $u = \cos x$, $v = \sin x$.

Step 3: Differentiate u and v with respect to x.

So $\dfrac{du}{dx} = -\sin x$ and $\dfrac{dv}{dx} = \cos x$

Recall:
$\cot x = \dfrac{1}{\tan x}$ (Section 2.2) and
$\tan x = \dfrac{\sin x}{\cos x}$ (C2 Section 3.6).

Step 4: Apply the quotient rule

$$\frac{dy}{dx} = \frac{v \dfrac{du}{dx} - u \dfrac{dv}{dx}}{v^2}$$

$$= \frac{\sin x \times (-\sin x) - \cos x \times \cos x}{\sin^2 x}$$

$$= \frac{-\sin^2 x - \cos^2 x}{\sin^2 x}$$

Step 5: Simplify.

$$= -\frac{(\sin^2 x + \cos^2 x)}{\sin^2 x}$$

$$= -\frac{1}{\sin^2 x}$$

$$= -\text{cosec}^2 x$$

Recall:
$\sin^2 x + \cos^2 x = 1$
(C2 Section 3.6).

Recall:
$\text{cosec}\, x = \dfrac{1}{\sin x}$ (Section 2.2).

1 Using the product rule, find $\dfrac{dy}{dx}$ when

a $y = x^3(x + 4)^2$ **b** $y = x \tan 3x$ **c** $y = e^{2x}x^4$

d $y = x \ln\sqrt{x + 3}$ **e** $y = e^{3x} \sin x$ **f** $y = x \sin 2x$

2 Using the quotient rule, find $\dfrac{dy}{dx}$ when

a $y = \dfrac{3x^2}{x - 3}$ **b** $y = \dfrac{\cos x}{1 - \sin x}$ **c** $y = \dfrac{e^{\frac{1}{2}x}}{2x^3}$

d $y = \dfrac{\ln (x + 1)}{x + 1}$ **e** $y = \dfrac{2x + 1}{3x - 2}$ **f** $y = \dfrac{\cos x}{x}$

3 Find f$'(x)$ when

a $f(x) = (x^2 + 3)^3(5x - 4)^5$ **b** $f(x) = e^x \cos^3 x$ **c** $f(x) = \dfrac{2e^x - 1}{2e^x + 1}$

d $f(x) = \dfrac{\sin 2x}{x^2}$ **e** $f(x) = x^2 \ln x$ **f** $f(x) = \dfrac{x^2}{e^x}$

4 The curve C has the equation $y = \dfrac{\ln x}{x^2}$. Find the gradient of C at the point where $x = $ e.

5 Find the coordinates of the stationary points on the following curves:

a $y = 2xe^x$ **b** $y = x^3 e^{-x}$

 6 It is given that $f(x) = \dfrac{x^2}{x - 3}$.

     a Show that $f'(x) = \dfrac{x^2 - 6x}{(x - 3)^2}$.   **b** Find $f''(x)$ in its simplest form.

7 Find an equation of the normal to the curve $y = \dfrac{x^2}{1 + x^2}$ at the point where $x = 1$, giving your answer in the form $ax + by = c$.

8 The curve C has equation $y = 2x^{\frac{1}{2}}e^{-x}$. Find the x-coordinate of the stationary point of the curve.

9 a Given that $y = x\sqrt{1 + x}$, show that $\dfrac{dy}{dx} = \dfrac{2 + 3x}{2\sqrt{1 + x}}$.

    b Given that $y = \dfrac{x}{1 + 2x}$, show that $\dfrac{dy}{dx} = \dfrac{1}{(1 + 2x)^2}$.

10 Use the derivatives of $\sin x$ and $\cos x$ to prove that the derivative of $\tan x$ is $\sec^2 x$.

11 It is given that $y = e^{\sin x}$. By differentiating, find

    a $\dfrac{dy}{dx}$ **b** $\dfrac{d^2y}{dx^2}$

SKILLS CHECK **4B EXTRA is on the CD**

Differentiation by the use of $\dfrac{dy}{dx} = \dfrac{1}{\dfrac{dx}{dy}}$.

When a function is defined in terms of y instead of x, a variation of the chain rule enables the derivative of y with respect to x to be found using

$$\frac{dy}{dx} = \frac{1}{\dfrac{dx}{dy}}.$$

So, if x is given as a function of y, differentiate x with respect to y to find $\dfrac{dx}{dy}$ and then work out the reciprocal to get $\dfrac{dy}{dx}$.

Example 4.15 Given that $x = \sin 3y$, show that $\dfrac{dy}{dx} = \frac{1}{3} \sec 3y$.

Step 1: Differentiate x with respect to y.

$x = \sin 3y$

$$\frac{dx}{dy} = 3 \cos 3y$$

Recall:
$\dfrac{d}{dx}(\sin ky) = k \cos ky$
(Section 4.2).

Step 2: Work out the reciprocal to find $\dfrac{dy}{dx}$.

$$\frac{dy}{dx} = \frac{1}{\dfrac{dx}{dy}}$$

$$= \frac{1}{3 \cos 3y}$$

Recall:
$\dfrac{1}{\cos y} = \sec y$ (Section 2.2).

Step 3: Write the result in the required format.

$$= \frac{1}{3} \times \frac{1}{\cos 3y}$$

$$= \tfrac{1}{3} \sec 3y$$

Note:
Usually questions will expect the answer in terms of y, since the original function was defined that way, but watch out for questions asking for the answer in terms of x.

Example 4.16 Consider the function $y = \ln x$.

Express x in terms of y and hence show that $\dfrac{d}{dx}(\ln x) = \dfrac{1}{x}$.

Step 1: Express x in terms of y.

$y = \ln x$

$\Rightarrow \quad x = e^y$

Recall:
e^x is the inverse of $\ln x$
(Section 3.2).

Step 2: Differentiate x with respect to y.

$$\frac{dx}{dy} = e^y$$

Step 3: Rewrite in terms of x.

$$= x$$

Step 4: Work out the reciprocal to find $\dfrac{dy}{dx}$.

$$\frac{dy}{dx} = \frac{1}{\dfrac{dx}{dy}} = \frac{1}{x}$$

i.e. $\quad \dfrac{d}{dx}(\ln x) = \dfrac{1}{x}$

1 Given that $x = \dfrac{y^2 - 5}{2}$, find $\dfrac{dy}{dx}$ in terms of y.

2 Given that $x = \sin y$, find

    a $\dfrac{dy}{dx}$ in terms of y **b** $\dfrac{dy}{dx}$ in terms of x

3 Given that $x = \dfrac{y^2 + y}{y - 1}$, find $\dfrac{dy}{dx}$ in terms of y.

4 Given that $x = y^4$, find $\dfrac{dy}{dx}$ in terms of x.

5 A curve has equation $x = \dfrac{y^2 + 2}{6}$.

    a Find $\dfrac{dy}{dx}$.

    b Find an equation of the normal to the curve at the point where $y = 1$, giving your answer in the form $ax + by = c$.

6 The curve C has equation $x = \dfrac{y}{e^y}$. Find the gradient of the tangent to C at the point where $y = \ln 3$.

7 Given the curve $x = \dfrac{y^2}{4}$, where $x > 0$, $y > 0$,

    a find equations for the tangents to the curve at $x = 1$ and at $x = 4$,

    b find the coordinates of the point of intersection of these two tangents.

8 Given the curve with equation $x = \sqrt{y^2 + 25}$, find an equation of the tangent to the curve at the point $(13, 12)$.

9 Consider the function $y = 2^x$.

    a Express x in the form $a \ln y$, stating the exact value of a.

    b Show that $\dfrac{d}{dx}(2^x) = \ln 2(2^x)$.

SKILLS CHECK **4C EXTRA** is on the CD

Examination practice 4: Differentiation

1 The graph of $y = \ln x - 3x$ has one stationary point.

    a Find $\dfrac{dy}{dx}$ and $\dfrac{d^2y}{dx^2}$.

    b Find the x-coordinate of the stationary point.

    c Find the value of $\dfrac{d^2y}{dx^2}$ at the stationary point and hence state whether the stationary point is a maximum or a minimum.

                                                          [AQA Jan 2003]

2 The curve $y = \frac{1}{2}\ln(3x + 5)$ crosses the axes at $P(0, p)$ and $Q(q, 0)$.

    a Find the exact values of p and q.

    b Find an equation of the normal to the curve at Q, giving your answer in the form $ay + bx + c = 0$, where a, b and c are integers.

3 A curve has equation $y = (x^2 + 5x + 4)\cos 3x$.

    a Find $\dfrac{dy}{dx}$.

    b Find the equation of the tangent to the curve at the point where $x = 0$.   [AQA June 2002]

4 A curve has the equation $y = \dfrac{2x}{\sin x}$, $0 < x < \pi$.

    a Find $\dfrac{dy}{dx}$.

    b The point P on the curve has coordinates $\left(\dfrac{\pi}{2}, \pi\right)$.

        i Show that the equation of the tangent to the curve at P is $y = 2x$.

        ii Find the equation of the normal to the curve at P, giving your answer in the form $y = mx + c$.   [AQA June 2003]

5 A curve has equation $y = \dfrac{x - 2}{x^2 + 5}$.

    a Determine the x-coordinates of the stationary points of the curve.

    b Find the equation of the tangent to the curve at the point where $x = 2$.   [AQA June 2001]

6 A curve has equation $y = x^2 - 3x + \ln x + 2$, $x > 0$.

    a **i** Find $\dfrac{dy}{dx}$.

        ii Hence show that the gradient of the curve at the point where $x = 2$ is $\frac{3}{2}$.

    b **i** Show that the x-coordinates of the stationary points of the curve satisfy the equation $2x^2 - 3x + 1 = 0$.

        ii Hence find the x-coordinates of each of the stationary points.

        iii Find $\dfrac{d^2y}{dx^2}$.

        iv Find the value of $\dfrac{d^2y}{dx^2}$ at each of the stationary points.

        v Hence show that the y-coordinate of the maximum point is $\frac{3}{4} - \ln 2$.   [AQA May 2002]

 7 Given that $f(x) = \dfrac{x^2 + 3x + 2}{x + 3}$ where $x < -3$,

    a find $f'(x)$,

    b hence solve the equation $f'(x) = \frac{7}{8}$.

 8 $f(x) = e^x \ln x$.

    a Find $f'(x)$

    b Find the value of $f''(1)$, giving your answer exactly.

9 $f(x) = \ln(\cos 3x)$.

    a Given that $f'(x) = a \tan 3x$, where a is an integer, find the value of a.

    b Find $f''(x)$, writing your answer in terms of $\tan 3x$.

10 a The function f is defined for $-\dfrac{\pi}{6} < x < \dfrac{\pi}{6}$ by $f(x) = \tan 3x$.
      The graph of $y = f(x)$ is sketched below.

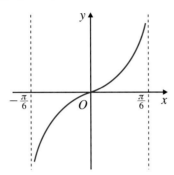

      The inverse of f is f^{-1}.

      i Sketch the graph of $y = f^{-1}(x)$.

      ii Find an expression for $f^{-1}(x)$.

    b A curve has equation $x = \tan 3y$.

      i Find $\dfrac{dy}{dx}$ in terms of y.

      ii Hence find the value of $\dfrac{dy}{dx}$ when $y = \dfrac{\pi}{9}$.   [AQA Jan 2005]

11 Differentiate $y = \sin^3 x$ with respect to x, writing your answer in terms of $\cos x$.

12 Find $\dfrac{dy}{dx}$ in terms of x:

    a $y = \dfrac{1}{(2x - 1)^3}$ **b** $y = e^{3\cos x}$ **c** $\ln y = 2x$

13 A curve is defined for $x \geq -2$ by the equation $y = (x^2 + 3)\sqrt{(x + 2)}$.

    a Show that $\dfrac{dy}{dx} = 0$ when $x = -1$ and find the x-coordinate of the other stationary point.

    b Find the value of $\dfrac{d^2y}{dx^2}$ when $x = -1$. Hence determine whether the turning point when $x = -1$
      is a maximum or minimum point.   [AQA (AEB) Jan 2000]

5 Integration

5.1 Integration of standard functions

Integrate e^x, $\dfrac{1}{x}$, $\sin x$, $\cos x$.

In *Core 1* and *Core 2* you integrated **powers of x** using the formula

$$\int ax^n \, dx = \frac{a}{n+1} x^{n+1} + c, \text{ provided } n \neq -1$$

You also used the relationship

$$\int (f(x) \pm g(x)) \, dx = \int f(x) \, dx \pm \int g(x) \, dx$$

In *Core 3* the functions are extended to include the following. These are not given in the formulae booklet, so must be learnt.

$f(x)$	$\int f(x)\, dx$
e^x	$e^x + c$
$\dfrac{1}{x}$	$\ln\lvert x \rvert + c$
$\cos x$	$\sin x + c$
$\sin x$	$-\cos x + c$

> **Recall:**
> Integration (C1 Chapter 4, C2 Chapter 6).

> **Recall:**
> Increase the power of x by 1 and divide by this new power.

> **Recall:**
> You can integrate sums and differences term by term.

> **Note:**
> It is possible to incorporate the constant into the log expression by letting $c = \ln k$. Then
> $$\int \frac{1}{x} \, dx = \ln\lvert x \rvert + \ln k = \ln k \lvert x \rvert.$$

> **Note:**
> When integrating trig functions, radians must be used.

Example 5.1 Find

a $\displaystyle\int (x^2 - \sin x + 2e^x) \, dx$　　**b** $\displaystyle\int \left(\frac{2}{x} + \frac{1}{3x^2}\right) dx$

Step 1: Integrate term by term.

a $\displaystyle\int (x^2 - \sin x + 2e^x) \, dx = \tfrac{1}{3}x^3 - (-\cos x) + 2e^x + c$
$$= \tfrac{1}{3}x^3 + \cos x + 2e^x + c$$

Step 2: Write terms in index form, where appropriate.

b $\displaystyle\int \left(\frac{2}{x} + \frac{1}{3x^2}\right) dx = \int \left(\frac{2}{x} + \frac{1}{3}x^{-2}\right) dx$

Step 3: Integrate term by term.

$$= 2\ln\lvert x \rvert + \frac{1}{3} \times \frac{x^{-1}}{-1} + c$$

$$= 2\ln\lvert x \rvert + \frac{1}{3} \times \left(-\frac{1}{x}\right) + c$$

$$= 2\ln\lvert x \rvert - \frac{1}{3x} + c$$

> **Tip:**
> Take care with the signs.

> **Note:**
> In index form, $\dfrac{2}{x} = 2x^{-1}$, but this format is not helpful, since the formula for integrating x^n does not apply when $n = -1$.

Example 5.2 Evaluate $\displaystyle\int_e^{e^2} \frac{1}{2x} \, dx$.

Step 1: Integrate.

$$\int_e^{e^2} \frac{1}{2x} \, dx = \tfrac{1}{2} \int_e^{e^2} \frac{1}{x} \, dx$$

$$= \tfrac{1}{2}\left[\ln\lvert x \rvert\right]_e^{e^2}$$

$$= \tfrac{1}{2}(\ln e^2 - \ln e)$$

Step 2: Substitute the limits and evaluate.

$$= \tfrac{1}{2}(2\ln e - \ln e)$$

$$= \tfrac{1}{2}\ln e$$

$$= \tfrac{1}{2}$$

> **Recall:**
> Definite integration (C1 Section 4.3).

> **Tip:**
> Take out the numerical factor.

> **Recall:**
> Substitute the upper limit, then subtract the result when the lower limit is substituted.

> **Recall:**
> $\ln e = 1$ (Section 3.2).

Example 5.3 Given that $\int_{-4}^{-2}\left(x + \dfrac{5}{x}\right)dx = p + q \ln 2$, where p and q are integers, find the value of p and the value of q.

Step 1: Integrate term by term.

Step 2: Substitute the limits and evaluate.

$$\int_{-4}^{-2}\left(x + \frac{5}{x}\right)dx = \left[\tfrac{1}{2}x^2 + 5\ln|x|\right]_{-4}^{-2}$$

$$= \tfrac{1}{2}\times(-2)^2 + 5\ln|-2| - \left(\tfrac{1}{2}\times(-4)^2 + 5\ln|-4|\right)$$

$$= 2 + 5\ln 2 - (8 + 5\ln 4)$$

$$= -6 + 5\ln\tfrac{2}{4}$$

$$= -6 + 5\ln 2^{-1}$$

$$= -6 - 5\ln 2$$

Step 3: Compare with the given result.

Hence $p = -6$ and $q = -5$.

Note:
The modulus function is very important here.

Tip:
If $a > 0$, $\ln|-a| = \ln a$

Recall:
$\log a - \log b = \log\frac{a}{b}$ and $\log a^n = n\log a$ (C2 Section 4.2).

Example 5.4 The diagram shows a sketch of the curve $y = 2 - e^x$. The curve crosses the x-axis at P. The region R, bounded by the curve and the coordinate axes, is shown shaded in the diagram.

Recall:
Area under curve (C1 Section 4.4, C2 Section 6.2).

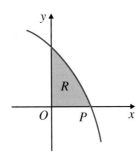

a Find the x-coordinate of P, giving the *exact* value.

b The area of R is $\ln a + b$, where a and b are integers. Find the values of a and b.

Step 1: Set $y = 0$ and solve.

a At P,
$$y = 0$$
$$\Rightarrow \quad 2 - e^x = 0$$
$$e^x = 2$$
$$x = \ln 2$$

The x-coordinate of P is $\ln 2$.

Recall:
$e^a = b \Leftrightarrow a = \ln b$ (Section 3.2).

Tip:
Give the exact value by leaving it in terms of the natural logarithm.

Step 2: Integrate using the area formula.

b $$\int_a^b y\,dx = \int_0^{\ln 2}(2 - e^x)\,dx$$

$$= \left[2x - e^x\right]_0^{\ln 2}$$

Step 3: Substitute the limits and evaluate.

$$= 2\ln 2 - e^{\ln 2} - (0 - e^0)$$

$$= 2\ln 2 - 2 + 1$$

$$= \ln 4 - 1$$

Step 4: Write in the required form and compare values to state a and b.

Hence $a = 4$ and $b = -1$.

Recall:
$e^{\ln a} = a$,
$e^0 = 1$,
$n\ln a = \ln a^n$
(Section 3.2).

5.2 Integration by inspection and by substitution

Simple cases of integration by inspection or substitution or by substitution (the reverse process of the chain rule).

Integration by inspection

In Section 4.2 you applied the chain rule to differentiate standard functions. Reversing this process leads to some standard integrals which, if you recognise them, could save you time. The basic ones are shown in the table below.

Tip:
Recognising the reverse of the chain rule to help you integrate is called 'integrating by inspection'.

$f(x)$	$\int f(x)\,dx$
e^{kx}	$\dfrac{1}{k}e^{kx} + c$
$\cos kx$	$\dfrac{1}{k}\sin kx + c$
$\sin kx$	$-\dfrac{1}{k}\cos kx + c$
$\sec^2 kx$	$\dfrac{1}{k}\tan kx + c$

Note:
The last result is given in the formulae booklet.

Example 5.5 Find

a $\displaystyle\int (2e^{5x} - \sin 3x)\,dx$ **b** $\displaystyle\int 6\sec^2 2x\,dx$

Tip:
Take care with fractions and signs.

Step 1: Integrate term by term, by inspection.

a $\displaystyle\int (2e^{5x} - \sin 3x)\,dx = 2 \times \tfrac{1}{5}e^{5x} - \left(-\tfrac{1}{3}\cos 3x\right) + c$

$$= \tfrac{2}{5}e^{5x} + \tfrac{1}{3}\cos 3x + c$$

Tip:
Remember to include the integration constant.

Step 2: Integrate by inspection.

b $\displaystyle\int 6\sec^2 2x\,dx = 6 \times \tfrac{1}{2}\tan 2x + c$

$$= 3\tan 2x + c$$

Integration by inspection can be applied more generally to several types of integrals. In the following type, look for a function raised to a power, multiplied by the derivative, or a numerical multiple of the derivative, of the function.

$$\int [f(x)]^n \times f'(x)\,dx = \frac{1}{n+1}[f(x)]^{n+1} + c, \quad n \neq -1$$

Note:
Sometimes you will need a numerical 'adjustment' factor. This is illustrated in the examples below.

Example 5.6 Find

a $\displaystyle\int x(1 + x^2)^5\,dx$ **b** $\displaystyle\int x^2\sqrt{2 + x^3}\,dx$ **c** $\displaystyle\int \frac{2x}{(1 - x^2)^3}\,dx$

General method:
Step 1: Check whether it can be integrated by inspection by differentiating the function being raised to a power.

a $\displaystyle\int x(1 + x^2)^5\,dx = \tfrac{1}{2} \times \tfrac{1}{6}(1 + x^2)^6 + c$

$$= \tfrac{1}{12}(1 + x^2)^6 + c$$

Side working:
$$\frac{d}{dx}(1 + x^2) = 2x$$
Factor of $\tfrac{1}{2}$ is needed.

Step 2: Calculate the numerical adjustment factor if required.

b $\int x^2 \sqrt{2 + x^3}\, dx = \int x^2 (2 + x^3)^{\frac{1}{2}}\, dx$

$\qquad = \frac{1}{3} \times \frac{1}{\frac{3}{2}} (2 + x^3)^{\frac{3}{2}} + c$

$\qquad = \frac{2}{9} (2 + x^3)^{\frac{3}{2}} + c$

Side working:

$\dfrac{d}{dx}(2 + x^3) = 3x^2$

Factor of $\frac{1}{3}$ is needed.

c $\int \dfrac{2x}{(1 - x^2)^3}\, dx = \int 2x(1 - x^2)^{-3}\, dx$

$\qquad = (-1) \times \dfrac{1}{-2}(1 - x^2)^{-2} + c$

$\qquad = \frac{1}{2}(1 - x^2)^{-2} + c$

Side working:

$\dfrac{d}{dx}(1 - x^2) = -2x$

Factor of -1 is needed.

A *special case* of this type is when you have to integrate a linear function of x, raised to a power. In this case:

$$\int (ax + b)^n\, dx = \dfrac{1}{a(n + 1)}(ax + b)^{n + 1} + c, \text{ provided } n \neq -1$$

For example

$$\int (2x + 3)^5\, dx = \dfrac{1}{2 \times 6}(2x + 3)^6 + c = \dfrac{1}{12}(2x + 3)^6 + c$$

In a question which can be integrated by inspection, you may be given a prompt in the examination, as in the following example:

Example 5.7 **a** Find $\dfrac{dy}{dx}$ when $y = (x^3 + 8)^6$.

b Hence, or otherwise, find $\int x^2(x^3 + 8)^5\, dx$.

a $y = (x^3 + 8)^6$

Step 1: Define t as a function of x; y is now a function of t.

Let $t = x^3 + 8$, then $y = t^6$

So $\dfrac{dt}{dx} = 3x^2$

Step 2: Differentiate t with respect to x, and y with respect to t.

and $\dfrac{dy}{dt} = 6t^5$

$\qquad = 6(x^3 + 8)^5$

Step 3: Rewrite t in terms of x.

Step 4: Apply the chain rule.

By the chain rule

$\dfrac{dy}{dx} = \dfrac{dy}{dt} \times \dfrac{dt}{dx}$

$\qquad = 6(x^3 + 8)^5 \times 3x^2$

$\qquad = 18x^2(x^3 + 8)^5$

b From part **a**

Step 1: Compare the integral with the answer in part **a** and adjust using a numerical factor.

$\int 18x^2(x^3 + 8)^5\, dx = (x^3 + 8)^6 + k$

Hence $\int x^2(x^3 + 8)^5\, dx = \frac{1}{18}(x^3 + 8)^6 + c$

Another type that can be integrated by inspection is e raised to a function of x and multiplied by the derivative, or a numerical multiple of the derivative, of that function of x.

$$\int e^{f(x)} \times f'(x)\, dx = e^{f(x)} + c$$

Example 5.8 Find

$$\textbf{a} \quad \int e^{3x+2}\, dx \qquad\qquad \textbf{b} \quad \int 3x^2\, e^{x^3}\, dx$$

General method:
Step 1: Check whether it can be integrated by inspection by differentiating $e^{f(x)}$.

Step 2: Calculate the numerical adjustment factor if required.

a $\quad \int e^{3x+2}\, dx = \tfrac{1}{3} e^{3x+2} + c$

Side working:
$$\frac{d}{dx}(e^{3x+2}) = 3e^{3x+2}$$
Factor of $\tfrac{1}{3}$ is needed.

b $\quad \int 3x^2\, e^{x^3}\, dx = e^{x^3} + c$

Side working: $\dfrac{d}{dx}(e^{x^3}) = 3x^2 e^{x^3}$
No adjustment factor is needed.

> **Tip:**
> When f(x) is linear
> $$\int e^{ax+b}\, dx = \frac{1}{a} e^{ax+b} + c.$$

An important example of integration by inspection involves looking for a fractional expression where the numerator is the derivative, or a multiple of the derivative, of the denominator. In this case

$$\int \frac{f'(x)}{f(x)}\, dx = \ln|f(x)| + c$$

Example 5.9 Find

$$\textbf{a} \quad \int \frac{1}{2+3x}\, dx \qquad \textbf{b} \quad \int \frac{\cos x}{1+2\sin x}\, dx \qquad \textbf{c} \quad \int \frac{e^{-x}}{1-e^{-x}}\, dx$$

General method:
Step 1: Check whether it can be integrated by inspection by differentiating the denominator.

Step 2: Calculate the numerical adjustment factor if required.

a $\quad \int \dfrac{1}{2+3x}\, dx = \tfrac{1}{3} \ln|2+3x| + c$

Side working:
$$\frac{d}{dx}(2+3x) = 3$$
Factor of $\tfrac{1}{3}$ is needed.

b $\quad \int \dfrac{\cos x}{1+2\sin x}\, dx = \tfrac{1}{2} \ln|1+2\sin x| + c$

Side working:
$$\frac{d}{dx}(1+2\sin x) = 2\cos x$$
Factor of $\tfrac{1}{2}$ is needed.

c $\quad \int \dfrac{e^{-x}}{1-e^{-x}}\, dx = \ln|1-e^{-x}| + c$

Side working:
$$\frac{d}{dx}(1-e^{-x}) = e^{-x}$$
No adjustment factor is needed.

> **Tip:**
> When f(x) is linear
> $$\int \frac{1}{ax+b}\, dx = \frac{1}{a} \ln|ax+b| + c.$$

Example 5.10 **a** By expressing $\tan x$ in terms of $\sin x$ and $\cos x$, show that

$$\int \tan x\, dx = \ln|\sec x| + c$$

b By expressing $\cot x$ in terms of $\sin x$ and $\cos x$, find $\displaystyle\int \cot x\, dx$.

Step 1: Use an appropriate trig identity.

a $\quad \tan x = \dfrac{\sin x}{\cos x} \quad\Rightarrow\quad \displaystyle\int \tan x\, dx = \int \dfrac{\sin x}{\cos x}\, dx$

$$= -\ln|\cos x| + c$$
$$= \ln|(\cos x)^{-1}| + c$$
$$= \ln|\sec x| + c$$

> **Tip:**
> Since $\dfrac{d}{dx}(\cos x) = -\sin x$, spot the use of $\displaystyle\int \frac{f'(x)}{f(x)}\, dx = \ln|f(x)| + c$

> **Tip:**
> $n \log x = \log x^n$

Step 2: Integrate recognising an appropriate standard result.

b $\cot x = \dfrac{\cos x}{\sin x} \Rightarrow \displaystyle\int \cot x \, dx = \int \dfrac{\cos x}{\sin x} \, dx$

$$= \ln |\sin x| + c$$

Tip:
$\dfrac{d}{dx}(\sin x) = \cos x$, so the numerator is the derivative of the denominator.

In fact, the results in Example 5.10 are printed in the formulae booklet and can be quoted if the derivation is not requested. These and similar results are given as follows:

$f(x)$	$\int f(x) \, dx$
$\tan x$	$\ln |\sec x| + c$
$\cot x$	$\ln |\sin x| + c$
$\mathrm{cosec}\, x$	$-\ln |\mathrm{cosec}\, x + \cot x| + c$
$\sec x$	$\ln |\sec x + \tan x| + c$

Example 5.11 Find $\displaystyle\int (1 + \tan x)^2 \, dx$.

Step 1: Expand the bracket.
Step 2: Express $\tan^2 x$ in terms of $\sec^2 x$.
Step 3: Integrate, using standard integrals.

$\displaystyle\int (1 + \tan x)^2 \, dx = \int (1 + 2\tan x + \tan^2 x) \, dx$

$\qquad\qquad\qquad\quad = \displaystyle\int (2\tan x + \sec^2 x) \, dx$

$\qquad\qquad\qquad\quad = 2\ln |\sec x| + \tan x + c$

Tip:
$1 + \tan^2 x \equiv \sec^2 x$ (Section 2.3).

Recall:
$\displaystyle\int \sec^2 x \, dx = \tan x + c$

Integration by substitution

Using an appropriate **substitution** enables some functions to be integrated. The method uses the reverse of the chain rule.

Using the substitution u, where u is a function of x,

$$y = \int f(x) \, dx$$

becomes

$$y = \int f(x) \dfrac{dx}{du} \, du$$

Note:
The method will always work on examples that could have been integrated by inspection.

Recall:
The chain rule (Section 4.2).

Example 5.12 Use the substitution $u = 2x - 1$ to find $\displaystyle\int (2x - 1)^8 \, dx$.

Step 1: Use the substitution to find $\dfrac{dx}{du}$.

Let $u = 2x - 1$, then $\dfrac{du}{dx} = 2 \Rightarrow \dfrac{dx}{du} = \dfrac{1}{2}$

Step 2: Rewrite the integral so that it is with respect to u.

$\displaystyle\int (2x - 1)^8 \, dx = \int (2x - 1)^8 \dfrac{dx}{du} \, du$

$\qquad\qquad\qquad = \displaystyle\int u^8 \times \tfrac{1}{2} \, du$

Step 3: Substitute to get an integral in terms of u.
Step 4: Integrate with respect to u.
Step 5: Rewrite in terms of x.

$\qquad\qquad\qquad = \tfrac{1}{2} \displaystyle\int u^8 \, du$

$\qquad\qquad\qquad = \tfrac{1}{2} \times \tfrac{1}{9} u^9 + c$

$\qquad\qquad\qquad = \tfrac{1}{18} (2x - 1)^9 + c$

Tip:
This could also be done by inspection.

Recall:
$\dfrac{dx}{du} = \dfrac{1}{\frac{du}{dx}}$ (Section 4.3).

Tip:
You could find $\dfrac{dx}{du}$ by making x the subject, where
$x = \dfrac{u + 1}{2} = \tfrac{1}{2}u + \tfrac{1}{2}$.

Example 5.13 Use the substitution $u = x^3 + 8$ to find $\int x^2(x^3 + 8)^5 \, dx$.

Note:
This is an alternative method for Example 5.7.

Step 1: Use the substitution to find $\dfrac{dx}{du}$.

Let $u = x^3 + 8$, then $\dfrac{du}{dx} = 3x^2 \Rightarrow \dfrac{dx}{du} = \dfrac{1}{3x^2} \Rightarrow x^2\dfrac{dx}{du} = \dfrac{1}{3}$

Step 2: Rewrite the integral so that it is with respect to u.

$\int x^2(x^3 + 8)^5 \, dx = \int x^2(x^3 + 8)^5 \dfrac{dx}{du} \, du$

Tip:
Substitute for $x^2\dfrac{dx}{du}$ in the integral.

$= \int (x^3 + 8)^5 x^2 \dfrac{dx}{du} \, du$

Step 3: Substitute to get an integral in terms of u.

$= \int u^5 \times \tfrac{1}{3} \, du$

$= \tfrac{1}{3}\int u^5 \, du$

Step 4: Integrate with respect to u.
Step 5: Rewrite in terms of x.

$= \tfrac{1}{3} \times \tfrac{1}{6} u^6 + c$

$= \tfrac{1}{18}(x^3 + 8)^6 + c$

Example 5.14 Use the substitution $u = 2x + 1$ to find $\int x\sqrt{2x + 1} \, dx$.

Tip:
This could not be done by inspection.

Step 1: Use the substitution to find x and $\dfrac{dx}{du}$.

Let $u = 2x + 1$, then $\dfrac{du}{dx} = 2 \Rightarrow \dfrac{dx}{du} = \dfrac{1}{2}$

Note:
You need to substitute for x so write x in terms of u.

Also $x = \dfrac{u - 1}{2} = \tfrac{1}{2}(u - 1)$

Step 2: Rewrite the integral so that it is with respect to u.

$\int x\sqrt{2x + 1} \, dx = \int x\sqrt{2x + 1} \dfrac{dx}{du} \, du$

Step 3: Substitute to get an integral in terms of u.

$= \int \tfrac{1}{2}(u - 1)u^{\frac{1}{2}} \times \tfrac{1}{2} \, du$

$= \tfrac{1}{4}\int (u^{\frac{3}{2}} - u^{\frac{1}{2}}) \, du$

Step 4: Integrate with respect to u.

$= \tfrac{1}{4} \times \left(\dfrac{1}{\frac{5}{2}} u^{\frac{5}{2}} - \dfrac{1}{\frac{3}{2}} u^{\frac{3}{2}} \right) + c$

Step 5: Rewrite in terms of x.

$= \tfrac{1}{4} \times (\tfrac{2}{5}(2x + 1)^{\frac{5}{2}} - \tfrac{2}{3}(2x + 1)^{\frac{3}{2}}) + c$

$= \tfrac{1}{10}(2x + 1)^{\frac{5}{2}} - \tfrac{1}{6}(2x + 1)^{\frac{3}{2}} + c$

Definite integration using a substitution

When performing definite integration using a substitution, it is advisable to change the limits to the limits of the new variable. This is illustrated in the following example.

Example 5.15 By using the substitution $u = 3x + 1$, show that

$$\int_1^2 \dfrac{x}{3x + 1} \, dx = p + q \ln r$$

where p, q and r are positive rational numbers to be found.

Step 1: Rewrite the integral so that it is with respect to u.

Let $u = 3x + 1$

$\int_1^2 \dfrac{x}{3x + 1} \, dx = \int_{x=1}^{x=2} \dfrac{x}{3x + 1} \dfrac{dx}{du} \, du$

Step 2: Substitute for x, $\dfrac{dx}{du}$ and the new limits.

$= \int_{u=4}^{u=7} \dfrac{\tfrac{1}{3}(u - 1)}{u} \times \dfrac{1}{3} \, du$

Side working:
$u = 3x + 1$
$\dfrac{du}{dx} = 3 \Rightarrow \dfrac{dx}{du} = \dfrac{1}{3}$
$x = \dfrac{u - 1}{3} = \tfrac{1}{3}(u - 1)$

Limits:

x	1	2
u	4	7

58

Step 3: Integrate with respect to u.	$= \frac{1}{9}\int_4^7 \left(1 - \frac{1}{u}\right) du$
	$= \frac{1}{9}\big[u - \ln|u|\big]_4^7$
Step 5: Substitute the limits and evaluate.	$= \frac{1}{9}(7 - \ln 7 - (4 - \ln 4))$
	$= \frac{1}{9}(7 - \ln 7 - 4 + \ln 4)$
Step 5: Express in the required format and state the values of p, q and r.	$= \frac{1}{9}\left(3 + \ln \frac{4}{7}\right)$
	$= \frac{1}{3} + \frac{1}{9}\ln \frac{4}{7}$

Hence $p = \frac{1}{3}$, $q = \frac{1}{9}$ and $r = \frac{4}{7}$.

Example 5.16 **a** Use the substitution $u = \sin x$ to find $\int \sin^3 x \cos x \, dx$.

b Hence evaluate $\int_0^{\frac{1}{2}\pi} \sin^3 x \cos x \, dx$.

Tip:
Exam questions often split the question into two parts like this.

Step 1: Use the substitution to find x and $\frac{dx}{du}$.	**a** Let $u = \sin x$, then $\dfrac{du}{dx} = \cos x \Rightarrow \cos x \dfrac{dx}{du} = 1$
Step 2: Rewrite the integral so that it is with respect to u.	$\int \sin^3 x \cos x \, dx = \int \sin^3 x \left(\cos x \dfrac{dx}{du}\right) du$
Step 3: Substitute to get an integral in terms of u.	$= \int u^3 \times 1 \, du$
Step 4: Integrate with respect to u.	$= \frac{1}{4}u^4 + c$
Step 5: Rewrite in terms of x.	$= \frac{1}{4}\sin^4 x + c$
Step 1: Using your answer to part **a**, substitute the limits and evaluate.	**b** $\int_0^{\frac{1}{2}\pi} \sin^3 x \cos x \, dx = \left[\frac{1}{4}\sin^4 x\right]_0^{\frac{1}{2}\pi}$
	$= \frac{1}{4}\sin^4\left(\frac{1}{2}\pi\right) - \frac{1}{4}\sin^4 0$
	$= \frac{1}{4}$

Tip:
Substitute for $\cos x \dfrac{dx}{du}$ in the integral.

Note:
When substituting the limits, the integration constant c is omitted.

Standard integrals involving a trigonometric substitution

Example 5.17 Use the substitution $x = 2\tan u$ to find $\int \dfrac{1}{4 + x^2} \, dx$.

Note:
x is a function of u, so you will get $\dfrac{dx}{du}$ straight away when you differentiate.

Step 1: Use the substitution to find $\frac{dx}{du}$.	Let $x = 2\tan u$, then $\dfrac{dx}{du} = 2\sec^2 u$.
Step 2: Simplify the expression in x.	Also, $4 + x^2 = 4 + 4\tan^2 u = 4(1 + \tan^2 u) = 4\sec^2 u$.
Step 3: Rewrite the integral so that it is with respect to u.	$\int \dfrac{1}{4 + x^2} \, dx = \int \dfrac{1}{4 + x^2} \dfrac{dx}{du} \, du$
Step 4: Substitute to get an integral in terms of u.	$= \int \dfrac{1}{4\sec^2 u} \times 2\sec^2 u \, du$
	$= \int \frac{1}{2} \, du$
Step 5: Integrate with respect to u.	$= \frac{1}{2}u + c$
Step 6: Rewrite in terms of x.	$x = 2\tan u \Rightarrow \tan u = \dfrac{x}{2} \Rightarrow u = \tan^{-1}\left(\dfrac{x}{2}\right)$
	So $\int \dfrac{1}{4 + x^2} \, dx = \frac{1}{2}\tan^{-1}\left(\dfrac{x}{2}\right) + c$.

Recall:
$1 + \tan^2 A \equiv \sec^2 A$

Recall:
Inverse trigonometric functions (Section 2.1).

The integral in Example 5.17 could have been found directly by using a standard result that can be quoted from your formulae booklet.

There are two results that you should know how to apply. These are as follows:

$f(x)$	$\int f(x)\,dx$
$\dfrac{1}{\sqrt{a^2 - x^2}}$	$\sin^{-1}\left(\dfrac{x}{a}\right) + c \quad (\lvert x \rvert < a)$
$\dfrac{1}{a^2 + x^2}$	$\dfrac{1}{a}\tan^{-1}\left(\dfrac{x}{a}\right) + c$

Note:
This has been derived using the substitution $x = a\sin u$.

Note:
This has been derived using the substitution $x = a\tan u$.

It may be necessary to take out a numerical factor before quoting the result.

Example 5.18   **a** $\displaystyle\int \frac{1}{\sqrt{9 - x^2}}\,dx$

          b   **i**   Write $9 - 4x^2$ in the form $4(a^2 - x^2)$.

             ii   Hence find $\displaystyle\int \frac{1}{\sqrt{9 - 4x^2}}\,dx$.

Step 1: Compare with an appropriate standard integral and quote the standard result.

a Comparing with the standard integral $\displaystyle\int \frac{1}{\sqrt{a^2 - x^2}}\,dx$:

$a^2 = 9 \Rightarrow a = 3$

So $\displaystyle\int \frac{1}{\sqrt{9 - x^2}}\,dx = \sin^{-1}\left(\frac{x}{3}\right) + c.$

Step 2: Equate coefficients to find a.

b   **i** $9 - 4x^2 = 4(a^2 - x^2)$

Equating the constant term gives

$9 = 4a^2$

$a^2 = \frac{9}{4} \Rightarrow a = \frac{3}{2}$

So $9 - 4x^2 = 4\left(\frac{9}{4} - x^2\right)$.

Step 3: Write the denominator using part **a**, taking out the numerical factor.

ii $\displaystyle\int \frac{1}{\sqrt{9 - 4x^2}}\,dx = \int \frac{1}{\sqrt{4\left(\frac{9}{4} - x^2\right)}}\,dx$

$= \displaystyle\int \frac{1}{2\sqrt{\left(\frac{9}{4} - x^2\right)}}\,dx$

Step 4: Compare with a standard integral and quote the standard result.

$= \frac{1}{2}\displaystyle\int \frac{1}{\sqrt{\left(\frac{9}{4} - x^2\right)}}\,dx$

$= \frac{1}{2}\sin^{-1}\left(\frac{2x}{3}\right) + c$

Tip:
$\sqrt{4\left(\frac{9}{4} - x^2\right)} = \sqrt{4} \times \sqrt{\frac{9}{4} - x^2}$ so remember to square root the 4.

Note:
$a = \frac{3}{2}$ so $\sin^{-1}\left(\frac{x}{a}\right)$
$= \sin^{-1}\left(x \div \frac{3}{2}\right) = \sin^{-1}\left(\frac{2x}{3}\right)$.

Example 5.19   Find the value of $\displaystyle\int_0^1 \frac{1}{5 + x^2}\,dx$, giving your answer to three significant figures.

Step 1: Compare with a standard integral and quote the standard result.

Comparing with the standard integral $\displaystyle\int \frac{1}{a^2 + x^2}\,dx$:

$a^2 = 5 \Rightarrow a = \sqrt{5}$

$\displaystyle\int_0^1 \frac{1}{5 + x^2}\,dx = \left[\frac{1}{\sqrt{5}}\tan^{-1}\left(\frac{x}{\sqrt{5}}\right)\right]_0^1$

Step 2: Substitute the limits and evaluate.

$= \frac{1}{\sqrt{5}}\left(\tan^{-1}\left(\frac{1}{\sqrt{5}}\right) - \tan^{-1}0\right)$

$= 0.1880\ldots$

$= 0.188 \text{ (3 s.f.)}$

Tip:
Don't think that this cannot be done as the square root of 5 isn't an integer. Leave it in surd form.

Tip:
You must work in radians. Make sure your calculator is in the correct mode.

Tip:
Give your answer to the required accuracy.

Example 5.20 **a** Find a, where $100 + 49x^2 = 49(a^2 + x^2)$.

b Hence find $\displaystyle\int \frac{1}{100 + 49x^2}\,dx$.

Step 1: Equate coefficients to find a.

a $100 + 49x^2 = 49(a^2 + x^2)$

Equating the constant term gives

$$100 = 49a^2$$
$$a^2 = \tfrac{100}{49}$$
$$a = \tfrac{10}{7}$$

Tip:
The coefficient of x^2 in the bracket is now 1.

Step 2: Write the denominator using part **a** and take out the numerical factor.
Step 3: Compare with a standard integral and quote the standard result.

b $\displaystyle\int \frac{1}{100 + 49x^2}\,dx = \int \frac{1}{49\left(\frac{100}{49} + x^2\right)}\,dx$

$$= \tfrac{1}{49} \int \frac{1}{\left(\frac{100}{49} + x^2\right)}\,dx$$

$$= \tfrac{1}{49} \times \tfrac{7}{10}\tan^{-1}\left(\frac{7x}{10}\right) + c$$

$$= \tfrac{1}{70}\tan^{-1}\left(\frac{7x}{10}\right) + c$$

Note:
$\dfrac{1}{a} = 1 \div \tfrac{10}{7} = \tfrac{7}{10}$ and
$\tan^{-1}\left(\dfrac{x}{a}\right) = \tan^{-1}\left(x \div \tfrac{10}{7}\right)$
$= \tan^{-1}\left(\dfrac{7x}{10}\right)$

SKILLS CHECK **5A: Integration by inspection and by substitution**

1 By inspection, or by using an appropriate substitution, find

a $\displaystyle\int e^{3x+1}\,dx$

b $\displaystyle\int -e^{-u}\,du$

c $\displaystyle\int \frac{1}{e^{2t}}\,dt$

d $\displaystyle\int \frac{1}{3(x+1)}\,dx$

e $\displaystyle\int \frac{2}{1+5x}\,dx$

f $\displaystyle\int \frac{3}{1-6x}\,dx$

g $\displaystyle\int \cos\tfrac{1}{3}x\,dx$

h $\displaystyle\int \frac{\sin 2y}{2}\,dy$

i $\displaystyle\int \frac{1}{3\sec 3x}\,dx$

2 Find the area of the region enclosed by the x-axis and the curve $y = \sin x$ where $0 \leqslant x \leqslant \pi$.

 3 The gradient at the point (x, y) on the curve $y = f(x)$ is given by $e^{3x} - 2x$. The curve goes through $(0, 1)$. Find an equation of the curve.

4

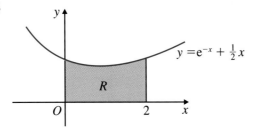

The diagram shows a sketch of the curve $y = e^{-x} + \tfrac{1}{2}x$. The region R (shown shaded) is bounded by the curve, the coordinate axes and the line $x = 2$.

Show that the area of region R is $\dfrac{2e^2 - 1}{e^2}$.

5 By inspection or by using an appropriate substitution, evaluate

a $\displaystyle\int_0^1 (3x-1)^5\,dx$

b $\displaystyle\int_{-\frac{1}{2}}^4 \sqrt{1+2x}\,dx$

c $\displaystyle\int_{-\frac{1}{4}\pi}^{\frac{1}{4}\pi} 3\cos 2x\,dx$

6 By inspection, or by using the substitution $u = x - 4$,

    a show that $\displaystyle\int_5^8 \frac{1}{x-4}\,dx = 2\ln 2$,

    b evaluate $\displaystyle\int_5^8 \frac{1}{(x-4)^2}\,dx$.

 7 Use the substitution $u = 2x + 1$ to show each of the following:

    a $\displaystyle\int x(2x+1)^3\,dx = \tfrac{1}{80}(8x-1)(2x+1)^4 + c$

    b $\displaystyle\int_0^1 \frac{x}{(2x+1)^3}\,dx = \frac{1}{18}$

8 Given that $\displaystyle\int_1^2 \left(\frac{2}{x} - 3x^2\right)dx = a + b\ln 2$, where a and b are integers, find the values of a and b.

9 By using the substitution $u = x^2 + 2$, or otherwise, find $\displaystyle\int xe^{x^2 + 2}\,dx$.

 10 Find

    a $\displaystyle\int \frac{3}{\sqrt{9-x^2}}\,dx$ **b** $\displaystyle\int \frac{3}{\sqrt{9-x}}\,dx$

11 Evaluate $\displaystyle\int_0^1 \frac{1}{x^2+1}\,dx$, giving your answer to three significant figures.

12 Using the substitution $u = e^{-x}$, or otherwise, find $\displaystyle\int \frac{e^{-x}}{1+e^{-x}}\,dx$.

13 Using the substitution $u = 1 + \cos x$, or otherwise, find $\displaystyle\int \frac{2\sin x}{1+\cos x}\,dx$.

14 a Using the substitution $u = 2x - 3$, find $\displaystyle\int x\sqrt{2x-3}\,dx$.

    b Using the substitution $u = 2 + x$, find $\displaystyle\int x(2+x)^6\,dx$.

15 a Using the substitution $u = \cos x$, or otherwise, find $\displaystyle\int \cos^2 x \sin x\,dx$.

    b Hence evaluate $\displaystyle\int_0^{\frac{1}{2}\pi} \cos^2 x \sin x\,dx$.

16 a i Write $1 - 25x^2$ in the form $25(a^2 - x^2)$.

    ii Hence find $\displaystyle\int \frac{1}{\sqrt{1-25x^2}}\,dx$.

    b i Write $9 + 25x^2$ in the form $25(a^2 + x^2)$.

    ii Hence find $\displaystyle\int \frac{1}{9+25x^2}\,dx$.

17 a Find $\displaystyle\int \frac{x+1}{4+x^2}\,dx$. $\left(\text{Hint: } \dfrac{x+1}{4+x^2} = \dfrac{x}{4+x^2} + \dfrac{1}{4+x^2}\right)$

    b Find $\displaystyle\int \frac{1-3x}{\sqrt{1-x^2}}\,dx$. $\left(\text{Hint: } \dfrac{1-3x}{\sqrt{1-x^2}} = \dfrac{1}{\sqrt{1-x^2}} - \dfrac{3x}{\sqrt{1-x^2}}\right)$

SKILLS CHECK **5A EXTRA is on the CD**

Integration by parts (the reverse process of the product rule).

The method known as **integrating by parts** is useful for integrating a **product** of two functions of x. One function is denoted by u and the other function is denoted by $\dfrac{dv}{dx}$, where

$$\int u \frac{dv}{dx}\,dx = uv - \int v \frac{du}{dx}\,dx$$

Tip:
Try this method when you have a product that cannot be integrated by inspection or substitution.

Note:
This formula has been obtained from the product rule for differentiating (Section 4.3).

Example 5.21 Use integration by parts to find the following:

  a $\displaystyle\int xe^{3x}\,dx$ **b** $\displaystyle\int x^2e^{3x}\,dx$

Step 1: Decide on the functions u and $\dfrac{dv}{dx}$, and find $\dfrac{du}{dx}$ and v.

a Let $u = x$, then $\dfrac{du}{dx} = 1$

Let $\dfrac{dv}{dx} = e^{3x}$, then $v = \displaystyle\int e^{3x}\,dx = \tfrac{1}{3}e^{3x}$

Note:
$\int e^{kx}\,dx = \dfrac{1}{k}e^{kx} + c$, but the integration constant is omitted at this stage.

Step 2: Apply the formula for integrating by parts.

$\displaystyle\int u \frac{dv}{dx}\,dx = uv - \int v \frac{du}{dx}\,dx$

So $\displaystyle\int xe^{3x}\,dx = x \times \tfrac{1}{3}e^{3x} - \int \tfrac{1}{3}e^{3x} \times 1\,dx$

$= \tfrac{1}{3}xe^{3x} - \tfrac{1}{3} \times \tfrac{1}{3}e^{3x} + c$

$= \tfrac{1}{3}xe^{3x} - \tfrac{1}{9}e^{3x} + c$

Note:
Include the integration constant at the final stage when finding an indefinite integral.

Step 3: Decide on the functions u and $\dfrac{dv}{dx}$, and find $\dfrac{du}{dx}$ and v.

b Let $u = x^2$, then $\dfrac{du}{dx} = 2x$

Let $\dfrac{dv}{dx} = e^{3x}$, then $v = \displaystyle\int e^{3x}\,dx = \tfrac{1}{3}e^{3x}$

$\displaystyle\int x^2e^{3x}\,dx = x^2 \times \tfrac{1}{3}e^{3x} - \int \tfrac{1}{3}e^{3x} \times 2x\,dx$

Step 4: Apply the formula for integrating by parts.

Step 5: Apply the formula again.

$= \tfrac{1}{3}x^2e^{3x} - \tfrac{2}{3}\displaystyle\int xe^{3x}\,dx$

$= \tfrac{1}{3}x^2e^{3x} - \tfrac{2}{3}(\tfrac{1}{3}xe^{3x} - \tfrac{1}{9}e^{3x}) + c$

$= \tfrac{1}{3}x^2e^{3x} - \tfrac{2}{9}xe^{3x} + \tfrac{2}{27}e^{3x} + c$

Note:
You have to apply integration by parts again.

Tip:
Notice the connection with part **a**. You can quote your result. Just put one integration constant at the end.

Example 5.22 **a** Find $\displaystyle\int(3x + 1)\sin 2x\,dx$.

  b Hence evaluate $\displaystyle\int_0^{\frac{1}{2}\pi}(3x + 1)\sin 2x\,dx$.

Note:
This is a common approach in a question, where you are asked to find an indefinite integral first, then to evaluate the definite integral.

Step 1: Decide on the functions u and $\dfrac{dv}{dx}$, and find $\dfrac{du}{dx}$ and v.

a Let $u = 3x + 1$, then $\dfrac{du}{dx} = 3$

Let $\dfrac{dv}{dx} = \sin 2x$, then $v = \displaystyle\int \sin 2x\,dx = -\tfrac{1}{2}\cos 2x$

Note:
$\int \sin kx\,dx = -\dfrac{1}{k}\cos kx + c$, but omit the integration constant at this stage.

Step 2: Apply the formula for integrating by parts and tidy up the expressions.

$\displaystyle\int(3x + 1)\sin 2x\,dx = (3x + 1) \times \left(-\tfrac{1}{2}\cos 2x\right) - \int\left(-\tfrac{1}{2}\cos 2x\right) \times 3\,dx$

$= -\tfrac{1}{2}(3x + 1)\cos 2x + \tfrac{3}{2}\displaystyle\int\cos 2x\,dx$

$= -\tfrac{1}{2}(3x + 1)\cos 2x + \tfrac{3}{2} \times \tfrac{1}{2}\sin 2x + c$

$= -\tfrac{1}{2}(3x + 1)\cos 2x + \tfrac{3}{4}\sin 2x + c$

Step 3: Apply the limits to the answer for the indefinite integral and evaluate.

b $\displaystyle\int_0^{\frac{1}{2}\pi} (3x + 1) \sin 2x \, dx$

$= \left[-\frac{1}{2}(3x + 1)\cos 2x + \frac{3}{4}\sin 2x \right]_0^{\frac{1}{2}\pi}$

$= -\frac{1}{2}\left(\frac{3}{2}\pi + 1\right)\cos \pi + \frac{3}{4}\sin \pi - \left(-\frac{1}{2}(3 \times 0 + 1)\cos 0 + \frac{3}{4}\sin 0\right)$

$= -\frac{1}{2}\left(\frac{3}{2}\pi + 1\right)(-1) + 0 - \left(-\frac{1}{2} \times 1 + 0\right)$

$= \frac{3}{4}\pi + \frac{1}{2} + \frac{1}{2}$

$= \frac{3}{4}\pi + 1$

Tip:
To evaluate this **definite** integral, use your answer to part **a**, omit c and substitute the limits.

When evaluating a **definite integral**, if you are not asked to find the indefinite integral first, you can substitute the limits as you go along, using the result

$$\int_a^b u\frac{dv}{dx}\,dx = [uv]_a^b - \int_a^b v\frac{du}{dx}\,dx$$

This is illustrated in the following example.

Tip:
If you use this method, you will need to take care; do not try to do too many things at once.

Example 5.23 Evaluate $\displaystyle\int_0^{\frac{1}{2}\pi} x \cos x \, dx$.

Step 1: Decide on the functions u and $\dfrac{dv}{dx}$, and find $\dfrac{du}{dx}$ and v.

Let $u = x$, then $\dfrac{du}{dx} = 1$

Let $\dfrac{dv}{dx} = \cos x$, then $v = \displaystyle\int \cos x \, dx = \sin x$

Step 2: Apply the formula for integrating by parts.

$\displaystyle\int_0^{\frac{1}{2}\pi} x \cos x \, dx = [x \sin x]_0^{\frac{1}{2}\pi} - \int_0^{\frac{1}{2}\pi} \sin x \times 1 \, dx$

Step 3: Substitute the limits for the first part.

$= \left(\frac{1}{2}\pi \sin \frac{1}{2}\pi - 0\right) - \displaystyle\int_0^{\frac{1}{2}\pi} \sin x \, dx$

Step 4: Integrate the second part.

$= \frac{1}{2}\pi - [-\cos x]_0^{\frac{1}{2}\pi}$

$= \frac{1}{2}\pi + [\cos x]_0^{\frac{1}{2}\pi}$

Step 5: Substitute the limits for the second part.

$= \frac{1}{2}\pi + (\cos \frac{1}{2}\pi - \cos 0)$

$= \frac{1}{2}\pi - 1$

Tip:
$\cos 0 = 1$
Do not forget to consider this.

Choice of u and $\dfrac{dv}{dx}$

If the product involves a polynomial in x, then this is usually taken to be u. However, the exception to this is when the other function involves $\ln x$. This is illustrated in the following example.

Tip:
If the polynomial involves x^2, then integration by parts needs to be carried out twice.

Example 5.24 Find

a $\displaystyle\int x \ln x \, dx$

b $\displaystyle\int \ln x \, dx$

Step 1: Write the integral with $\ln x$ first.

a $\displaystyle\int x \ln x \, dx = \int (\ln x) \times x \, dx$

Step 2: Decide on the functions u and $\dfrac{dv}{dx}$, and find $\dfrac{du}{dx}$ and v.

Let $u = \ln x$, then $\dfrac{du}{dx} = \dfrac{1}{x}$

Let $\dfrac{dv}{dx} = x$, then $v = \displaystyle\int x \, dx = \frac{1}{2}x^2$

Step 3: Apply the formula for integrating by parts and tidy up the expressions.

$$\int (\ln x) \times x \, dx = \ln x \times \tfrac{1}{2}x^2 - \int \tfrac{1}{2}x^2 \times \frac{1}{x} \, dx$$

$$= \tfrac{1}{2}x^2 \ln x - \tfrac{1}{2}\int x \, dx$$

$$= \tfrac{1}{2}x^2 \ln x - \tfrac{1}{2} \times \tfrac{1}{2}x^2 + c$$

$$= \tfrac{1}{2}x^2 \ln x - \tfrac{1}{4}x^2 + c$$

Note:
You now have to integrate a simple function of x.

Step 4: Write the integral as a product with $\ln x$ first.

b $\int \ln x \, dx = \int (\ln x) \times 1 \, dx$

Step 5: Decide on the functions u and $\dfrac{dv}{dx}$, and find $\dfrac{du}{dx}$ and v.

Let $u = \ln x$, then $\quad \dfrac{du}{dx} = \dfrac{1}{x}$

Let $\dfrac{dv}{dx} = 1$, then $\quad v = \int 1 \, dx = x$

Tip:
Set up the product by multiplying by 1.

Step 6: Apply the formula for integrating by parts and tidy up the expressions.

$$\int (\ln x) \times 1 \, dx = \ln x \times x - \int x \times \frac{1}{x} \, dx$$

$$= x \ln x - \int 1 \, dx$$

$$= x \ln x - x + c$$

Hence $\int \ln x \, dx = x \ln x - x + c$.

SKILLS CHECK **5B: Integration by parts**

 1 Use integration by parts to find

a $\int (2x + 3) \cos x \, dx$ **b** $\int x \cos(2x + 3) \, dx$

2 Use integration by parts to find

a $\int x e^{3x} \, dx$ **b** $\int e^{3x}(x - 1) \, dx$

3 Evaluate $\int_{2}^{3} x^2 \ln x \, dx$, giving your answer to three significant figures.

4 a Find $\int 5x \sin 2x \, dx$.

     b Hence show that $\int_{0}^{\frac{1}{2}\pi} x(1 + 5 \sin 2x) \, dx = \tfrac{1}{8}\pi^2 + \tfrac{5}{4}\pi$.

5 a Use integration by parts to find $\int x e^{-2x} \, dx$.

     b

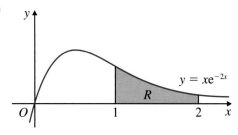

The diagram shows a sketch of the curve $y = x e^{-2x}$. Find the area of the region R, bounded by the curve, the x-axis and the lines $x = 1$ and $x = 2$.

6 a Find $\displaystyle\int_0^\pi x \cos x \, dx$.

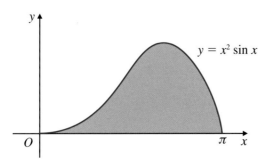

$y = x^2 \sin x$

The diagram shows a sketch of the curve $y = x^2 \sin x$ for values of x in the interval $0 \leqslant x \leqslant \pi$.

b Find the area of the region bounded by the curve and the x-axis.

 7 Using integration by parts, find the exact value, in terms of e, of $\displaystyle\int_1^e \frac{\ln x}{x^3} \, dx$.

8 Find the value of $\displaystyle\int_0^1 \frac{x^2}{e^x} \, dx$, giving your answer exactly in terms of e.

SKILLS CHECK **5B EXTRA** is on the CD

5.4 Volume of revolution

Evaluation of volume of revolution.

When a region is rotated completely about the x-axis or the y-axis, a solid is formed. The volume of this solid is known as the **volume of revolution**.

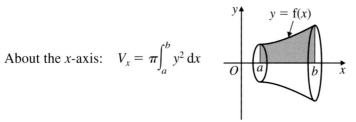

About the x-axis: $\displaystyle V_x = \pi \int_a^b y^2 \, dx$

$y = f(x)$

About the y-axis: $\displaystyle V_y = \pi \int_c^d x^2 \, dy$

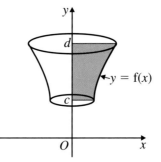

$y = f(x)$

> **TIP:**
> You should learn these formulae.

> **TIP:**
> Always note carefully which is the axis of rotation.

There are several ways of describing the rotation, such as:

- rotated completely
- rotated through 360°
- rotated through 2π radians
- rotated through four right angles.

Example 5.25 The diagram shows the curve $y = \dfrac{1}{2x+1}$.

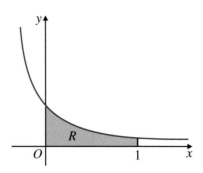

The region R (shown shaded in the diagram) is enclosed by the curve, the axes and the line $x = 1$.

The region R is rotated completely about the x-axis. Find the exact value of the volume of the solid formed.

Step 1: Apply the volume formula.

$$V_x = \pi \int_a^b y^2 \, dx$$

$$= \pi \int_0^1 \left(\frac{1}{2x+1} \right)^2 dx$$

$$= \pi \int_0^1 (2x+1)^{-2} \, dx$$

Step 2: Integrate with respect to x.

$$= \pi \left[\frac{1}{2 \times (-1)} (2x+1)^{-1} \right]_0^1$$

Step 3: Substitute the limits and evaluate.

$$= -\tfrac{1}{2}\pi \left[(2x+1)^{-1} \right]_0^1$$

$$= -\tfrac{1}{2}\pi (3^{-1} - 1^{-1})$$

$$= -\tfrac{1}{2}\pi \left(\tfrac{1}{3} - 1 \right)$$

$$= \tfrac{1}{3}\pi$$

Tip:
Remember to include π.

Tip:
Square first, then integrate.

Tip:
Write the expression in index form.

Tip:
$\int (ax+b)^n \, dx$
$= \dfrac{1}{a(n+1)}(ax+b)^{n+1}, n \neq -1$
If you prefer, substitute $u = 2x+1$.

Tip:
You are asked for the exact value, so leave your answer in terms of π.

Example 5.26 The diagram shows the curve $y = \ln 2x$.

The region bounded by the curve, the y-axis and the lines $y = 1$ and $y = 3$ (shown shaded), is rotated completely about the y-axis. Show that the volume of the solid formed is $\tfrac{1}{8}\pi e^2(e^2-1)(e^2+1)$.

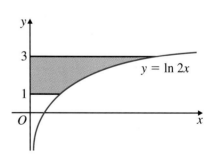

Step 1: Express x in terms of y.

$y = \ln 2x \quad \Rightarrow 2x = e^y$
$\qquad\qquad\qquad x = \tfrac{1}{2} e^y$

Step 2: Apply the volume formula.

$$V_y = \pi \int_c^d x^2 \, dy$$

$$= \pi \int_1^3 \left(\tfrac{1}{2} e^y \right)^2 dy$$

$$= \tfrac{1}{4}\pi \int_1^3 e^{2y} \, dy$$

Step 3: Integrate with respect to y.
Step 4: Substitute the limits and simplify to the required form.

$$= \tfrac{1}{4}\pi \left[\tfrac{1}{2} e^{2y} \right]_1^3$$

$$= \tfrac{1}{8}\pi (e^6 - e^2)$$

$$= \tfrac{1}{8}\pi e^2(e^4 - 1)$$

$$= \tfrac{1}{8}\pi e^2(e^2 - 1)(e^2 + 1)$$

Recall:
$a = \ln b \Leftrightarrow b = e^a$

Tip:
$(e^a)^2 = e^{2a}$. Also remember to square the $\tfrac{1}{2}$.

Tip:
Do not round using your calculator.

Tip:
Factorise the difference between two squares.

Example 5.27 The curves $y = x^2$ and $y = \sqrt{x}$ are shown in the sketch.

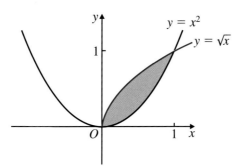

They intersect at $(0, 0)$ and $(1, 1)$. The region enclosed between the curves, shown shaded, is rotated through four right angles about the x-axis.

Note:
This means that it is rotated completely (through 360°).

Show that the volume of the solid formed is $\frac{3}{10}\pi$.

Step 1: Find the volume when the area 'under' each curve is rotated about the x-axis.

Consider the area under $y = \sqrt{x}$:

$$V_1 = \pi \int_a^b y^2 \, dx$$

$$= \pi \int_0^1 x \, dx$$

$$= \pi \left[\tfrac{1}{2} x^2 \right]_0^1$$

$$= \tfrac{1}{2} \pi (1 - 0)$$

$$= \tfrac{1}{2} \pi$$

Consider the area under $y = x^2$:

$$V_2 = \pi \int_a^b y^2 \, dx$$

$$= \pi \int_0^1 x^4 \, dx$$

$$= \pi \left[\tfrac{1}{5} x^5 \right]_0^1$$

$$= \tfrac{1}{5} \pi (1 - 0)$$

$$= \tfrac{1}{5} \pi$$

Step 2: Subtract the volumes.

The required volume is $V_1 - V_2 = \tfrac{1}{2}\pi - \tfrac{1}{5}\pi = \tfrac{3}{10}\pi$.

Alternatively, you could work out

$$\pi \int_a^b (y_1^2 - y_2^2) \, dx, \text{ provided that } y_1 > y_2 \text{ for } a \leqslant x \leqslant b.$$

SKILLS CHECK **5C: Volume of revolution**

 1 The diagram shows a sketch of the curve $y = \dfrac{1}{x}$, $x > 0$.

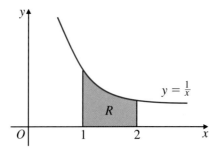

The region R is bounded by the x-axis and the lines $x = 1$ and $x = 2$.

a Find the exact value of the area of R.

b Find the volume of the solid formed when R is rotated completely about the x-axis.

2 The region bounded by the curve $y = e^x$, the coordinate axes and the line $x = 1$ is rotated completely about the x-axis. Find the exact volume of the solid formed.

68

3 The region bounded by the curve $y = x^3$, the y-axis and the lines $y = 1$ and $y = 8$ is rotated through $360°$ about the y-axis. Find the exact volume of the solid formed.

4 The diagram shows a sketch of the curve $y = \dfrac{1}{\sqrt{x+1}}$, $x > -1$.

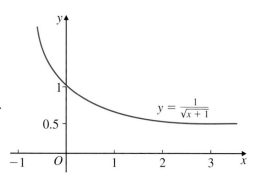

$y = \dfrac{1}{\sqrt{x+1}}$

a The region enclosed by the curve, the coordinate axes and the line $x = 1$ is rotated completely about the x-axis. Find the exact volume of the solid formed.

b The region enclosed by the curve, the y-axis and the lines $y = 0.5$ and $y = 1$ is rotated completely about the y-axis. Find the exact volume of the solid formed.

5 The diagram shows the curve $y = x^2 + 1$ and the line $y = x + 1$. The line and the curve intersect at A and B.

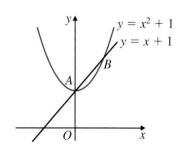

a Find the coordinates of A and B.

b The region enclosed between the curve and the line is rotated completely about the x-axis. Find the exact volume of the solid formed.

 6 The diagram shows the curve $y = x(4 - x)$ and the line $y = x$.

The line and the curve intersect at the origin and at P.

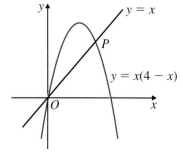

a Find the coordinates of P.

b The region enclosed between the curve and the line is rotated completely about the x-axis. Show that the volume of the solid formed is $\dfrac{108}{5}\pi$.

7 The region enclosed by the curves $y = x^2$ and $y = \sqrt{x}$ is rotated completely about the y-axis (see Example 5.27 for the sketch). Find the exact volume of the solid formed.

8 a Sketch the graph of $y = 9 - x^2$, indicating the coordinates of the points where the graph crosses the coordinate axes.

b The region between the curve and the x-axis from $x = -2$ to $x = -1$ is rotated through $360°$ about the x-axis. Find the exact volume of the solid generated.

c The region between the curve, the positive x-axis and the positive y-axis is rotated through $360°$ about the y-axis. Find the exact volume of the solid generated.

9 a i Find $\displaystyle\int_0^{\frac{1}{2}\pi} x \sin x \, dx.$ **ii** Find $\displaystyle\int_0^{\frac{1}{2}\pi} x^2 \cos x \, dx.$

b The diagram shows the curve $y = x\sqrt{\cos x}$ for $0 \leqslant x \leqslant \frac{1}{2}\pi$.

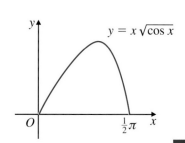

$y = x\sqrt{\cos x}$

The region bounded by the curve and the x-axis is rotated through $360°$ about the x-axis.
Find the exact volume of the solid generated.

10 a Use integration by parts to find $\int x\,e^{-2x}\,dx$.

b

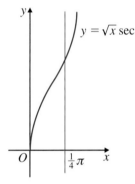

The diagram shows the curve $y = x\,e^{-x}$. The region R, bounded by the curve, the x-axis and the lines $x = 1$ and $x = 2$, is rotated completely about the x-axis.

Show that the volume of the solid generated is $\dfrac{\pi(5e^2 - 13)}{4e^4}$.

11 a Use integration by parts to find $\int x \sec^2 x\,dx$.

b

The diagram shows part of the curve of $y = \sqrt{x}\,\sec x$ and the line $x = \frac{1}{4}\pi$. The region enclosed by the curve, the line and the x-axis is rotated completely about the x-axis.

Show that the volume of the solid generated is $\frac{1}{4}\pi^2 - \frac{1}{2}\pi\ln 2$, given that $\cos\frac{1}{4}\pi = \dfrac{1}{\sqrt{2}}$.

SKILLS CHECK **5C EXTRA** is on the CD

Examination practice 5: Integration

1 Show that $\displaystyle\int_0^1 (e^{2x} + 4)\,dx = \frac{1}{2}(e^2 + 7)$. [AQA Jan 2004]

2 a Differentiate:

   i $\ln x$;   **ii** $\dfrac{1}{x}$.

b By expanding the bracket and integrating, show that

$$\int_1^9 \left(1 + \frac{3}{x}\right)^2 dx = 4(4 + 3\ln 3)$$ [AQA Nov 2003]

3 It is given that

$$f(x) = 2x^3 + 3x^2 + 7$$

a Find the derivative $f'(x)$, factorising your answer.

b Hence, show that $\displaystyle\int_0^2 \frac{x(x+1)}{f(x)}\,dx = k\ln 5$, stating the value of the constant k. [AQA June 2003]

4 Evaluate $\int_0^1 e^{2x} + x^{\frac{1}{2}} \, dx$, giving your answer in the form $pe^2 + q$, where p and q are rational numbers.

[AQA June 2001]

5 Use integration by parts to find $\int_0^{\frac{1}{2}} x \, e^{2x} \, dx$.

[AQA June 2003]

6 The diagram shows a sketch of the curve
$y = e^{2x} - 2$.

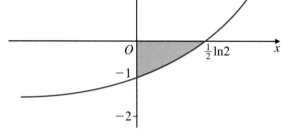

  a On separate diagrams, sketch the graphs of
the following curves showing the coordinates
of the points where the graph intersects the
coordinate axes:

  i $y = |e^{2x} - 2|$;   **ii** $y = e^{2x} - 5$.

  b **i** Find $\int (e^{2x} - 2) \, dx$.

   ii Hence show that the area of the shaded region bounded by the curve $y = e^{2x} - 2$ and the
coordinate axes is $\ln 2 - \frac{1}{2}$.

[AQA May 2002]

7 Evaluate $\int_0^{\frac{1}{3}\pi} \sin 3x \, dx$.

8 Use integration by parts to find $\int_0^{\pi} x^2 \sin x \, dx$.

9 **a** Differentiate $(3x + 4)^{\frac{1}{2}}$ with respect to x.

  b The region R is bounded by the curve with equation $y = \dfrac{1}{\sqrt{3x + 4}}$, the coordinate axes and the

   straight line with equation $x = 7$. Calculate the area of R.

[AQA (AEB) Jan 2000]

 10 Use the substitution $u = x + 1$ to express the integral

$$I = \int_0^1 \frac{x^2}{(x + 1)^2} \, dx$$

in the form

$$\int_a^b \left(1 - \frac{c}{u} + \frac{1}{u^2}\right) du,$$

stating the value of each of the constants a, b and c.

Hence find the exact value of I.

[AQA (AEB) Jan 2000]

11 **a** Use integration by parts to find $\int x^3 \ln x \, dx$.

  b By means of the substitution $u = 3 + e^x$, or otherwise, evaluate

$$\int_0^{\ln 6} e^{2x}(3 + e^x)^{\frac{1}{2}} \, dx.$$

[AQA (AEB) June 2000]

12 Evaluate $\int_0^1 \frac{1}{(2x + 1)^4} \, dx$, giving your answer as a fraction in its simplest form.

13 Find:

   a $\displaystyle\int e^{3-5x}\,dx$

   b $\displaystyle\int \frac{1}{4-3x}\,dx$

14 a Find   **i** $\displaystyle\int \frac{x}{9+x^2}\,dx$ **ii** $\displaystyle\int \frac{1}{9+x^2}\,dx.$

   b Given that $\displaystyle\int_0^1 \frac{2x+3}{9+x^2}\,dx = \ln a + \tan^{-1} b$, where a and b are positive rational numbers, use your answers from part **a**, or otherwise, to find the values of a and b.

15 The curve with equation

$$y = x\sqrt{x^2+3}$$

is sketched below. The region R, shaded on the diagram, is bounded by the curve, the x-axis and the line $x = 1$.

   a The region R is rotated through 2π radians about the x-axis. Find the volume of the solid generated.

   b Determine an equation of the tangent to the curve at the point where $x = 1$.

   c **i** Differentiate $(x^2+3)^{\frac{3}{2}}$ with respect to x.

     ii Use the result from part **c i** to find the area of region R.   [AQA Jan 2002]

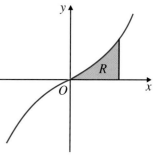

16 A curve has equation $y = 1 + \dfrac{6}{3x+2}$ and is sketched below for $0 \leq x \leq 2$.

The shaded region R is bounded by the curve, the coordinate axes and the line $x = 2$.

   a Express y^2 in the form $1 + \dfrac{A}{3x+2} + \dfrac{B}{(3x+2)^2}.$

   b Find:

     i $\displaystyle\int \frac{1}{3x+2}\,dx$;

     ii $\displaystyle\int \frac{1}{(3x+2)^2}\,dx.$

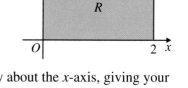

   c Find the volume of the solid formed when R is rotated completely about the x-axis, giving your answer to three significant figures.   [AQA Jan 2005]

17 Use the substitution $u = 1 + \sin x$, or otherwise, to determine

   a $\displaystyle\int \frac{\cos x}{1+\sin x}\,dx$ **b** $\displaystyle\int \frac{\cos x}{(1+\sin x)^2}\,dx$

18 By using the substitution $u = 3 + \ln x$, or otherwise, find $\displaystyle\int \frac{3+\ln x}{x}\,dx.$

6 Numerical methods

6.1 Location of roots

Location of roots of f(x) = 0 by considering changes of sign of f(x) in an interval of x in which f(x) is continuous.

If the function f is continuous in the interval $a \leqslant x \leqslant b$, and f($x$) changes sign in this interval, then f(x) = 0 has a root in the interval $a \leqslant x \leqslant b$.

Note:
The function **must** be continuous in the interval.

Note:
Use radians for trigonometric functions.

Example 6.1 It is given that f(x) = $\sin x - 2x + 1$.
Show that f(x) = 0 has a root in the interval $0.8 \leqslant x \leqslant 0.9$.

Step 1: Substitute the boundary values of the interval into f(x) and consider the signs.

f(0.8) = $\sin 0.8 - 2 \times 0.8 + 1 = 0.117\ldots > 0$

f(0.9) = $\sin 0.9 - 2 \times 0.9 + 1 = -0.016\ldots < 0$

Step 2: Look for a sign change.

Change of sign
$\Rightarrow$ root of f(x) = 0 lies in the interval $0.8 \leqslant x \leqslant 0.9$.

Tip:
Write down the value from your calculator to make it clear whether it is positive or negative.

Tip:
Be careful. The interval is $0.8 \leqslant x \leqslant 0.9$, **not** $0.8 \leqslant$ f(x) $\leqslant 0.9$.

Example 6.2 **a** Sketch, on the same axes, the graphs of $y = e^x$ and $y = 5 - x$.

Given that f(x) = $e^x + x - 5$,

b show that the equation f(x) = 0 has only one root,

c show that this root lies in the interval $1 < x < 2$.

Step 1: Sketch the graphs. **a**

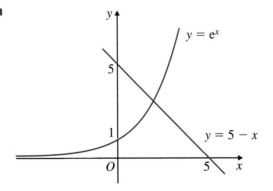

Recall:
Graph of e^x (Section 3.1).

Step 2: Consider the intersection point(s) of the two graphs.

b When the graphs intersect:

$$e^x = 5 - x$$
$$\Rightarrow \quad e^x + x - 5 = 0$$

From the sketch, it can be seen that the graphs of $y = 5 - x$ and $y = e^x$ have only one point of intersection

$\Rightarrow$ there is only one root of $e^x + x - 5 = 0$.

Hence the equation f(x) = 0 has only one root.

Tip:
Show your full method in case you make a slip.

Step 3: Substitute the boundary values of the interval into f(x) and consider the signs.

c f(x) $= e^x + x - 5$

f(1) $= e^1 + 1 - 5 = -1.28\ldots < 0$

f(2) $= e^2 + 2 - 5 = 4.38\ldots > 0$

Change of sign $\Rightarrow$ root of f(x) = 0 in the interval $1 < x < 2$.

Tip:
You must state 'change of sign' and make a comment about there being a root.

Approximate solutions of equations using simple iterative methods, including recurrence relations of the form $x_{n+1} = f(x_n)$.

An approximate solution of an equation can sometimes be found by rearranging the equation into the form $x_{n+1} = f(x_n)$ and then using this iterative formula, also known as a recurrence relation, with an appropriate starting value, to find subsequent approximations.

If $x_{n+1} = f(x_n)$ converges to a root α, then α is a root of the equation $x = f(x)$.

This method fails if the sequence of values does not converge.

Note:
A converging sequence approaches a limit.

The accuracy of an approximate solution can be tested by using the change of sign method. For example, if a solution of the equation $f(x) = 0$ is 4.327 to three decimal places, then $f(4.3265)$ and $f(4.3275)$ will have different signs.

Example 6.3 The sequence defined by the iterative formula

$$x_{n+1} = \sqrt[3]{7 - 3x_n},$$

with $x_1 = 1.4$, converges to α.

a Use the iterative formula to find α, to three decimal places. You should show the result of each iteration.

b State an equation of which α is a root.

Step 1: Apply the iterative formula for a few iterations.

a $\quad x_1 = 1.4$

$x_2 = \sqrt[3]{7 - 3(1.4)} = 1.4094\ldots$

$x_3 = \sqrt[3]{7 - 3(1.4094\ldots)} = 1.4046\ldots$

$x_4 = \sqrt[3]{7 - 3(1.4046\ldots)} = 1.4070\ldots$

$x_5 = \sqrt[3]{7 - 3(1.4070\ldots)} = 1.4058\ldots$

$x_6 = \sqrt[3]{7 - 3(1.4058\ldots)} = 1.4064\ldots$

So $\alpha = 1.406$ (3 d.p.)

Tip:
Use the full calculator display each time.

Tip:
You are told that the sequence converges, so if your values are not converging, check your working.

Step 2: State the value, to 3 d.p., to which the sequence appears to be converging.

Calculator note:
If your calculator has the facility of reproducing the last answer, try keying in the following:

$$\boxed{1.4} \quad \boxed{=} \quad \boxed{\sqrt[3]{}} \quad \boxed{(} \quad \boxed{7} \quad \boxed{-} \quad \boxed{3} \quad \boxed{\text{Ans}} \quad \boxed{)}$$

The successive terms of the iteration can then be obtained by keying in

$$\boxed{=} \quad \boxed{=} \quad \boxed{=} \quad \text{etc.}$$

Note:
On graphical calculators, press $\boxed{\text{EXE}}$ or $\boxed{\text{ENTER}}$

Step 3: Set up the equation $x = f(x)$.

b Since $x_{n+1} = f(x_n)$ converges to a root α, then α is a root of the equation $x = f(x)$.

So $\quad x = \sqrt[3]{7 - 3x}$

$\qquad x^3 = 7 - 3x$

$x^3 + 3x - 7 = 0$

Hence α is a root of the cubic equation $x^3 + 3x - 7 = 0$.

Example 6.4 It is given that $f(x) = \ln x + x + 4, x > 0$.

    a Show that $f(x) = 0$ has a root in the interval $0.01 < x < 0.02$.

    b Show that $f(x) = 0$ can be rearranged in the form $x = e^{-(x+4)}$.

    c Use the iteration $x_{n+1} = e^{-(x_n+4)}$, with $x_1 = 0.01$, to find the values of x_2, x_3 and x_4, giving the value of x_4 to four decimal places.

    d By considering the change of sign of $f(x)$ in a suitable interval, show that your value for x_4 gives an accurate estimate, correct to four decimal places, of the root of $f(x) = 0$.

Tip:
Even if you can't answer **b**, you could still do parts **c** and **d** as they are not dependent on previous answers.

Step 1: Substitute the boundary values of the interval into $f(x)$ and consider the signs.

a $f(x) = \ln x + x + 4$

     $f(0.01) = \ln 0.01 + 0.01 + 4 = -0.595... < 0$

     $f(0.02) = \ln 0.02 + 0.02 + 4 = 0.107... > 0$

     Change of sign $\Rightarrow$ root of $f(x) = 0$ in the given interval

Tip:
Always write down the value you get for your calculation to make it clear whether it is positive or negative.

Step 2: Rearrange the equation to the required format.

b $f(x) = 0$

     $\ln x + x + 4 = 0$

     $\ln x = -x - 4 = -(x + 4)$

     $\Rightarrow$ $x = e^{-(x+4)}$

Recall:
$\ln x = y \Leftrightarrow x = e^y$ (Section 3.2).

Step 3: Find the specified iterations.

c $x_{n+1} = e^{-(x_n+4)}$

     $x_1 = 0.01$

     $x_2 = e^{-(0.01+4)} = 0.01813...$

     $x_3 = e^{-(0.01813...+4)} = 0.01798...$

     $x_4 = e^{-(0.01798...+4)} = 0.01798... = 0.0180$ (4 d.p.)

Tip:
Since you are asked for the root to 4 d.p. you must work to at least 5 decimal places. Here the full calculator display has been used each time.

Calculator note:
Try keying in the following:

| 0.01 | | = | | e^x | | (−) | | (| | Ans | | + | | 4 | |) |

| = | | = | | = | etc.

Step 4: Consider the sign of $f(x)$ for appropriate values of x.

d $f(x) = \ln x + x + 4$

     $f(0.01795) = \ln (0.01795) + 0.01795 + 4 = -0.00221... < 0$

     $f(0.01805) = \ln (0.01805) + 0.01805 + 4 = 0.00344... > 0$

     Change of sign $\Rightarrow$ root of $f(x) = 0$ is between 0.01795 and 0.01805

     $\Rightarrow$ root is 0.0180 (correct to four decimal places)

Tip:
If the root is 0.0180, correct to four decimal places, then it must lie in the interval $0.01795 < x < 0.01805$.

Example 6.5 The sequence defined by the iterative formula

$$x_{n+1} = \sqrt{10 - 2x_n},$$

with $x_1 = 2$, converges to α.

    a Use the iterative formula for five iterations, giving the value of x_6 to three significant figures.

    b State an equation which has α as a root and hence find the exact value of α.

Step 1: Apply the iterative formula.

a $x_1 = 2$

$x_2 = \sqrt{10 - 2(2)} = 2.449\ldots$

$x_3 = \sqrt{10 - 2(2.449\ldots)} = 2.258\ldots$

$x_4 = \sqrt{10 - 2(2.258\ldots)} = 2.341\ldots$

$x_5 = \sqrt{10 - 2(2.341\ldots)} = 2.305\ldots$

$x_6 = \sqrt{10 - 2(2.305\ldots)} = 2.321\ldots = 2.32$ (3 s.f.)

Tip:
Show the result of each iteration.

Step 2: Set up the equation $x = f(x)$.

b $x_{n+1} = f(x_n)$ converges to a root α

$\Rightarrow$ α is a root of the equation $x = f(x)$

$\Rightarrow$ $x = \sqrt{10 - 2x}$

$x^2 = 10 - 2x$

$x^2 + 2x - 10 = 0$

Hence α is a root of the quadratic equation $x^2 + 2x - 10 = 0$.

Step 3: Solve the quadratic equation.

Completing the square:

$(x + 1)^2 - 1 = 10$

$(x + 1)^2 = 11$

$x + 1 = \pm\sqrt{11}$

$x = -1 \pm\sqrt{11}$

Tip:
You could use the quadratic formula, but completing the square works well here.

Tip:
Since the iteration converges to a positive value, choose the positive root.

Step 4: Choose the appropriate root.

Since $\alpha > 0$,

$\alpha = -1 +\sqrt{11}$

Note:
$-1 + \sqrt{11} = 2.32$ (3 s.f.), so the value in part **a** was a good approximation.

Divergence

Sometimes an iterative process does not lead to a root of the equation even if you use a starting value close to that root. The sequence x_1, x_2, x_3, … may diverge.

Example 6.6 **a** Show that the equation $\frac{1}{2}x^3 - 2x + 1 = 0$ has a root in the interval $1 < x < 2$.

b Use the iteration $x_{n+1} = \dfrac{x_n^3 + 2}{4}$ with $x_1 = 1.75$ to find x_2, x_3, x_4, x_5 and x_6, giving your answers to three decimal places.

c Comment on your sequence of values.

Step 1: Substitute the boundary values of the interval into $f(x)$ and consider the signs.

a Let $f(x) = \frac{1}{2}x^3 - 2x + 1$

$f(1) = \frac{1}{2}(1^3) - 2 \times 1 + 1 = -\frac{1}{2} < 0$

$f(2) = \frac{1}{2}(2^3) - 2 \times 2 + 1 = 1 > 0$

Change of sign $\Rightarrow$ root of $f(x) = 0$ in the interval $1 < x < 2$.

Tip:
Remember to give the reason.

Step 2: Apply the iterative formula.

b $x_{n+1} = \dfrac{x_n^3 + 2}{4}$

$x_1 = 1.75$

$x_2 = \dfrac{1.75^3 + 2}{4} = 1.8398\ldots = 1.840$ (3 d.p.)

$x_3 = \dfrac{1.8398\ldots^3 + 2}{4} = 2.0569\ldots = 2.057$ (3 d.p.)

$x_4 = \dfrac{2.0569\ldots^3 + 2}{4} = 2.6758\ldots = 2.676$ (3 d.p.)

$x_5 = \dfrac{2.6758\ldots^3 + 2}{4} = 5.2899\ldots = 5.290$ (3 d.p.)

$x_6 = \dfrac{5.2899\ldots^3 + 2}{4} = 37.5070\ldots = 37.507$ (3 d.p.)

Note:
The root is actually $x = 1.675$ (3 d.p.). Even though the iterative process started close to that root, the sequence is getting further and further away.

Calculator note:
Try keying in the following:

$\boxed{=}\;\boxed{=}\;\boxed{=}$ etc.

Tip:
If you are using a calculator to generate your iterations be careful with brackets and make sure you write down all your intermediate results.

Step 3: Comment on the sequence.

c The sequence diverges. The given iteration does not converge to a root of $\frac{1}{2}x^3 - 2x + 1 = 0$.

Note:
To find the root in the interval $1 < x < 2$ you would need a different iteration formula. Try $x = \sqrt[3]{4x - 2}$.

Cobweb and staircase diagrams

Iterations can be illustrated by **cobweb and staircase** diagrams and these are helpful in deciding whether the iteration converges or diverges.

For the iterative formula $x_{n+1} = f(x_n)$, draw the graphs of $y = f(x)$ and $y = x$.

Starting from x_1, draw a line vertically to the curve $y = f(x)$, then horizontally to $y = x$. The x-coordinate of this point gives x_2.

Repeat the procedure to get x_3, x_4, etc.

Here are some examples:

Cobweb diagrams:

Convergent

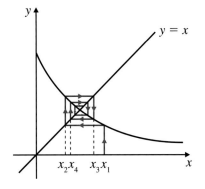

Divergent

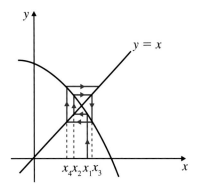

Staircase diagrams:

Convergent   Divergent

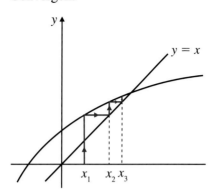

To illustrate the <u>iterations</u> in Example 6.5, consider the sketches of the graphs of $y = \sqrt{10 - 2x}$ and $y = x$, with $x_1 = 2$.

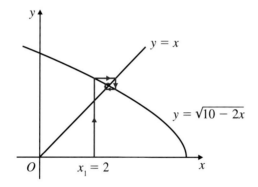

Note:
This cobweb diagram shows that, with the given starting value, the iteration converges.

To illustrate the iterations in Example 6.6, consider the sketches of the graphs of $y = \dfrac{x^3 + 2}{4}$ and $y = x$, with $x_1 = 1.75$.

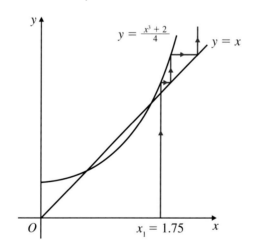

Note:
This staircase diagram shows that, with the given starting value, the iteration diverges.

SKILLS CHECK **6A: Approximate roots of equations**

1 Show that the equation $f(x) = 0$ has a root in the given interval.

a $f(x) = \sqrt[3]{x} + x - 7, 5 < x < 6$

b $f(x) = \cos 2x + x, -1 < x < 0$

c $f(x) = \ln(x - 4) + \sqrt{x}, 4.1 \leqslant x \leqslant 4.2$

d $f(x) = \tan x - e^x, -4 < x < -3$

e $f(x) = \dfrac{1}{x} + 1 - x^3, 1.2 < x < 1.3$

2 a On the same axes, sketch the graphs of $y = \ln x$ and $y = x^2 - 4$.

   b Hence write down the number of roots of the equation $\ln x - x^2 + 4 = 0$.

   c Show that the equation $\ln x - x^2 + 4 = 0$ has a root in the interval $2 \leqslant x \leqslant 3$.

3 a Using the same axes, sketch the graphs of $y = e^x - 1$ and $y = 2x + 1$.

   b Hence show that the equation $e^x - 2x - 2 = 0$ has one negative root and one positive root.

   The positive root of the equation $e^x - 2x - 2 = 0$ lies in the interval $n < x < n + 1$, where n is an integer.

   c Find the value of n.

4 In each of the following:

   i Show that the equation can be rearranged into the given iterative formula.

   ii Use the value of x_1 to find the values of x_2, x_3, x_4 and x_5, giving the value of x_5 to the stated accuracy.

   a $x^2 - \dfrac{2}{x} - 1 = 0$ $x_{n+1} = \sqrt{\dfrac{2}{x_n} + 1}$ $x_1 = 1.5$ 3 decimal places

   b $\cos x - 9x - 4 = 0$ $x_{n+1} = \frac{1}{9}(\cos x_n - 4)$ $x_1 = -0.3$ 5 decimal places

   c $\ln x + 2 - \sqrt{x} = 0$ $x_{n+1} = e^{\sqrt{x_n} - 2}$ $x_1 = 0.2$ 3 decimal places

   d $\tan x - 3x = 0$ $x_{n+1} = \tan^{-1}(3x_n)$ $x_1 = 1.32$ 3 decimal places

   e $e^{0.3x} - x - 2 = 0$ $x_{n+1} = \frac{10}{3} \ln (x_n + 2)$ $x_1 = 7.51$ 4 significant figures

5 The curve with equation $y = 4x^3 - x - 6$ intersects the x-axis at the point A where $x = \alpha$.

   a Show that α lies between 1 and 2.

   b Show that the equation $4x^3 - x - 6 = 0$ can be rearranged in the form $x = \sqrt[3]{\dfrac{x + 6}{4}}$.

   c Use the iterative formula $x_{n+1} = \sqrt[3]{\dfrac{x_n + 6}{4}}$ with $x_1 = 1$ to find x_4, giving your answer to four significant figures.

6 The sequence defined by the iteration $x_{n+1} = \ln(2x_n + 5)$ converges to α.

   a Use the iteration with $x_1 = 3$ to calculate x_3, giving your answer to three significant figures.

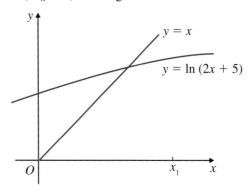

   b The diagram shows the graphs of $y = x$, and $y = \ln(2x + 5)$, and the position of x_1. Copy the diagram and draw a cobweb or staircase diagram to show how convergence takes place, indicating the positions of x_2 and x_3.

   c State an equation of which α is a root.

7 It is given that $f(x) = \cos x - x^2 + 3$.

   a Show that $f(x) = 0$ has a root in the interval $1 < x < 2$.

   b Using the iterative formula $x_{n+1} = \sqrt{\cos x_n + 3}$ and $x_1 = 1.7$, write down the values of x_2, x_3 and x_4, giving the value of x_4 to three decimal places.

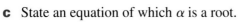

8   **a**   On the same axes, sketch the curves with equations $y = 2^x$ and $y = x^3 - 7$.

    b   Use your sketch to show that the equation $2^x - x^3 + 7 = 0$ has exactly one solution.

    c   Show that $2^x - x^3 + 7 = 0$ can be rearranged to $x = \sqrt[3]{2^x + 7}$.

    d   Use an appropriate iterative formula with $x_1 = 2.3$ to find x_2, x_3 and x_4.

    e   Hence write down an approximate solution of the equation $2^x - x^3 + 7 = 0$.

 9 $f(x) = x^2 - 4x - 8$.

    a   Show that $f(x) = 0$ has a root α in the interval $5 < x < 6$.

    b   **i**   Show that $f(x) = 0$ can be rearranged to the equation $x = \frac{1}{4}x^2 - 2$.

        ii   Use the iterative formula $x_{n+1} = \frac{1}{4}x_n^2 - 2$ and $x_1 = 5$ to find x_2, x_3, x_4, x_5 and x_6 and comment on your sequence of results.

    c   **i**   Show that $f(x) = 0$ can be rearranged to the equation $x = \sqrt{4x + 8}$.

        ii   Use the iterative formula $x_{n+1} = \sqrt{4x_n + 8}$ and $x_1 = 5$ to find x_2, x_3, x_4, x_5 and x_6 and comment on your sequence of results.

    d   State the value of α, to two decimal places.

    e   Compare this value with the *exact* root of $f(x) = 0$ in the interval $5 < x < 6$.

SKILLS CHECK   **6A EXTRA** is on the CD

6.3 Numerical integration of functions

Numerical integration of functions using the mid-ordinate rule and Simpson's rule.

Consider the region bounded by a curve $y = f(x)$, the x-axis and the lines $x = a$ and $x = b$.

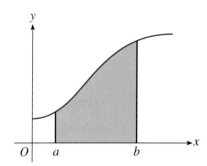

Recall:
Area under a curve
(C1 Section 4.4).

Recall:
If the region is above the x-axis, the integral gives a positive value, and if the region is below the x-axis, the integral gives a negative value.

Using integration, the exact area of the region is given by $\left| \int_a^b y \, dx \right|$.

If, however, this integral cannot be found, approximate methods for finding the area may be used. In *Core 2*, the trapezium rule was introduced. In *Core 3*, the methods are extended to include the following:

- the mid-ordinate rule
- Simpson's rule.

The mid-ordinate rule

To apply the mid-ordinate rule, the area is approximated by n rectangles, each of width h.

The height of each rectangle is the y-value halfway along each rectangle. This is known as the **mid-ordinate** and the mid-ordinates of all the rectangles are denoted by $y_{\frac{1}{2}}, y_{\frac{3}{2}}, y_{\frac{5}{2}}, \ldots, y_{n-\frac{1}{2}}$.

Tip:
y-values are called ordinates.

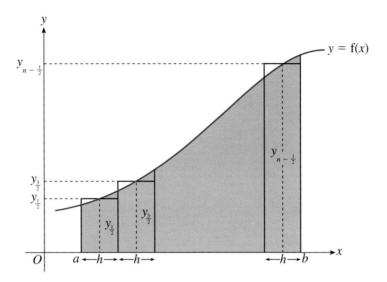

The mid-ordinate rule, which may be quoted, is

$$\int_a^b y \, dx = h(y_{\frac{1}{2}} + y_{\frac{3}{2}} + \cdots + y_{n-\frac{3}{2}} + y_{n-\frac{1}{2}}), \text{ where } h = \frac{b-a}{n}$$

Note:
Make sure that you know where this formula is given in the formulae booklet.

A better approximation can be obtained by increasing the number of rectangles.

Note:
This reduces the value of h.

Example 6.7 The diagram shows a sketch of the graph of $y = (\ln x)^2$.

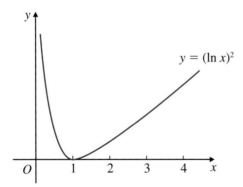

Giving your answer to three significant figures, estimate the value of $\int_2^4 (\ln x)^2 \, dx$ using the mid-ordinate rule with four strips of equal width.

Step 1: Calculate the value of h, the width of each strip.

$n = 4, a = 2, b = 4, \text{ so } h = \dfrac{4-2}{4} = 0.5$

	Mid *x*-value	Mid *y*-value i.e. mid-ordinate
$x_0 = 2$		
	$x_{\frac{1}{2}} = 2.25$	$y_{\frac{1}{2}} = (\ln 2.25)^2$
$x_1 = 2.5$		
	$x_{\frac{3}{2}} = 2.75$	$y_{\frac{3}{2}} = (\ln 2.75)^2$
$x_2 = 3$		
	$x_{\frac{5}{2}} = 3.25$	$y_{\frac{5}{2}} = (\ln 3.25)^2$
$x_3 = 3.5$		
	$x_{\frac{7}{2}} = 3.75$	$y_{\frac{7}{2}} = (\ln 3.75)^2$
$x_4 = 4$		

Tip:
It is a good idea to show your values for the mid-ordinates in a table. Here is a possible format.

Note:
$(\ln a)^2 \neq 2\ln a$.

Step 3: Apply the mid-ordinate rule.

$$\int_2^4 (\ln x)^2 \, \mathrm{d}x \approx h(y_{\frac{1}{2}} + y_{\frac{3}{2}} + y_{\frac{5}{2}} + y_{\frac{7}{2}})$$

$$= 0.5((\ln 2.25)^2 + (\ln 2.75)^2 + (\ln 3.25)^2 + (\ln 3.75)^2)$$

$$= 0.5 \times 4.8172\ldots$$

$$= 2.4086\ldots$$

$$= 2.41 \text{ (3 s.f.)}$$

Tip:
To avoid rounding errors, avoid using your calculator before the final stage.

Simpson's rule

To apply Simpson's rule, the region is split into *n* equal-width strips, where *n* is **even**. Again, the greater the number of strips, the better the approximation.

Note:
There is an odd number of *y*-values (ordinates).

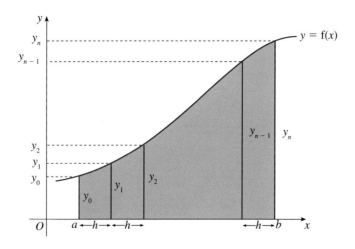

Simpson's rule, which may be quoted, is:

$$\int_a^b y \, \mathrm{d}x \approx \tfrac{1}{3} h\{(y_0 + y_n) + 4(y_1 + y_3 + \cdots + y_{n-1}) + 2(y_2 + y_4 + \cdots + y_{n-2})\}$$

where $h = \dfrac{b-a}{n}$ and *n* is even.

In words: $\int_a^b y \, \mathrm{d}x \approx \tfrac{1}{3} h\{\text{ends} + 4(\text{odds}) + 2(\text{evens})\}$

Note:
Simpson's rule is also given in the formulae booklet.

Example 6.8 Use Simpson's rule, with six strips of equal width, to find an approximate value for $\int_0^6 \sqrt{1 + x^3}\, dx$.

Step 1: Calculate h, the width of each strip.

There are 6 strips, so $n = 6$.

$n = 6$, $a = 0$, $b = 6$, so $h = \dfrac{b - a}{n} = \dfrac{6 - 0}{6} = 1$

Note:
There are 6 strips, so there are 7 ordinates.

Step 2: Work out the x-values and the appropriate y-values (ordinates).

	End ordinates	Odd ordinates	Even ordinates
$x_0 = 0$	$y_0 = \sqrt{1}$		
$x_1 = 1$		$y_1 = \sqrt{2}$	
$x_2 = 2$			$y_2 = \sqrt{9}$
$x_3 = 3$		$y_3 = \sqrt{28}$	
$x_4 = 4$			$y_4 = \sqrt{65}$
$x_5 = 5$		$y_5 = \sqrt{126}$	
$x_6 = 6$	$y_6 = \sqrt{217}$		

Tip:
Showing the ordinates in this format may help you to apply the formula accurately.

Step 3: Apply Simpson's rule and evaluate.

$\int_0^6 \sqrt{1 + x^3}\, dx \approx \frac{1}{3} h\{(y_0 + y_6) + 4(y_1 + y_3 + y_5) + 2(y_2 + y_4)\}$

$= \frac{1}{3} \times 1 \times \{(\sqrt{1} + \sqrt{217}) + 4(\sqrt{2} + \sqrt{28} + \sqrt{126}) + 2(\sqrt{9} + \sqrt{65})\}$

$= \frac{1}{3} \times 109.578\ldots$

$= 36.526\ldots$

$= 36.5$ (3 s.f.)

Tip:
To avoid rounding errors, avoid using your calculator before the final stage.

Example 6.9 Giving your answer to three significant figures, find an approximate value for $\int_0^1 \sin^{-1} x\, dx$

a using Simpson's rule with five ordinates (four strips),

b using the mid-ordinate rule with five strips.

Step 1: Calculate h, the width of each strip.

a There are 4 strips, so $n = 4$.

$n = 4$, $a = 0$, $b = 1$, so $h = \dfrac{b - a}{n} = \dfrac{1 - 0}{4} = 0.25$

Tip:
n is the number of strips.

Step 2: Work out the x-values and the appropriate y-values (ordinates).

	End ordinates	Odd ordinates	Even ordinates
$x_0 = 0$	$y_0 = \sin^{-1} 0$		
$x_1 = 0.25$		$y_1 = \sin^{-1} 0.25$	
$x_2 = 0.5$			$y_2 = \sin^{-1} 0.5$
$x_3 = 0.75$		$y_3 = \sin^{-1} 0.75$	
$x_4 = 1$	$y_4 = \sin^{-1} 1$		

Tip:
You must work in radians when using trig functions, so remember to set your calculator to radians mode.

Step 3: Apply Simpson's rule and evaluate.

$\int_0^1 \sin^{-1} x\, dx \approx \frac{1}{3} h\{(y_0 + y_4) + 4(y_1 + y_3) + 2(y_2)\}$

$= \frac{1}{3} \times 0.25 \times \{(\sin^{-1} 0 + \sin^{-1} 1) + 4(\sin^{-1} 0.25 + \sin^{-1} 0.75) + 2(\sin^{-1} 0.5)\}$

$= \frac{1}{3} \times 0.25 \times 7.02096\ldots$

$= 0.58508\ldots$

$= 0.585$ (3 s.f.)

Tip:
Work out the curly bracket first and write down its value. Then multiply by $\frac{1}{3} h$.

b There are 5 strips, so $n = 5$.

$n = 5$, $a = 0$, $b = 1$, so $h = \dfrac{b - a}{n} = \dfrac{1 - 0}{5} = 0.2$

Note:
Any number of strips may be chosen for the mid-ordinate rule, odd or even.

Step 5: Work out the appropriate mid-ordinate values.

	Mid x-value	Mid y-value i.e. mid-ordinate
$x_0 = 0$		
	$x_{\frac{1}{2}} = 0.1$	$y_{\frac{1}{2}} = \sin^{-1}0.1$
$x_1 = 0.2$		
	$x_{\frac{3}{2}} = 0.3$	$y_{\frac{3}{2}} = \sin^{-1}0.3$
$x_2 = 0.4$		
	$x_{\frac{5}{2}} = 0.5$	$y_{\frac{5}{2}} = \sin^{-1}0.5$
$x_3 = 0.6$		
	$x_{\frac{7}{2}} = 0.7$	$y_{\frac{7}{2}} = \sin^{-1}0.7$
$x_4 = 0.8$		
	$x_{\frac{9}{2}} = 0.9$	$y_{\frac{9}{2}} = \sin^{-1}0.9$
$x_5 = 1$		

Step 6: Apply the mid-ordinate rule.

$$\int_0^1 \sin^{-1} x \, dx \approx h(y_{\frac{1}{2}} + y_{\frac{3}{2}} + y_{\frac{5}{2}} + y_{\frac{7}{2}} + y_{\frac{9}{2}})$$

$$= 0.2(\sin^{-1}0.1 + \sin^{-1}0.3 + \sin^{-1}0.5 + \sin^{-1}0.7 + \sin^{-1}0.9)$$

$$= 0.2 \times 2.823\ldots$$

$$= 0.5647\ldots$$

$$= 0.565 \ (3 \text{ s.f.})$$

Tip:
Again, do not use your calculator until this final stage and remember to set it to radians mode.

It can be shown, by integrating, that the exact answer is $\sin^{-1}1 - 1 = 0.57079\ldots$, so both methods give reasonable approximations, with Simpson's rule overestimating the value and the mid-ordinate rule underestimating it.

Note:
Integrate the product $\sin^{-1} x \times 1$ using the integration by parts formula.

SKILLS CHECK **6B: Numerical integration**

1 Use the mid-ordinate rule with four strips of equal width to find an estimate of $\displaystyle\int_2^6 \dfrac{1}{x^2 - 1} \, dx$.

2 a Use the mid-ordinate rule with five strips of equal width to find an estimate of $\displaystyle\int_0^2 e^{3x} \, dx$, giving your answer to three significant figures.

b Find the exact value of $\displaystyle\int_0^2 e^{3x} \, dx$ by integrating.

c Find the percentage error in using your answer in part **a** for $\displaystyle\int_0^2 e^{3x} \, dx$.

3 Giving your answer to three significant figures, use the mid-ordinate rule with six strips of equal width to estimate the value of $\displaystyle\int_0^3 \sqrt[3]{1 + x^2} \, dx$.

4 Use Simpson's rule, with 7 ordinates (6 strips), to find an approximate value for $\int_1^7 \ln(x^2 + 2)\,dx$.

 5 a Use Simpson's rule, with four strips of equal width, to find an approximate value for $\int_0^2 e^{x^2}\,dx$.

    b Explain how to obtain a better approximation using Simpson's rule.

6 a Sketch the curve $y = \tan^{-1}x$.

    b Using Simpson's rule with 9 ordinates (8 strips), estimate the area of the region bounded by the curve $y = \tan^{-1}x$, the line $x = 1$ and the x-axis.

    c The exact value of the area is $\frac{1}{4}\pi - \frac{1}{2}\ln 2$. Compare your estimate with this value.

7 a Use Simpson's rule, with six strips, to find an approximate value for $\int_1^4 \frac{1}{1 + x^2}\,dx$.
Give your answer to four decimal places.

    b The diagram shows the curve $y = \dfrac{1}{\sqrt{1 + x^2}}$, $x \geq 0$.

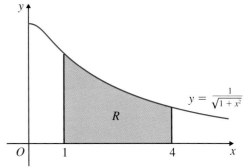

The region R bounded by the curve, the x-axis and the lines $x = 1$ and $x = 4$ is rotated completely about the x-axis.

      i Using Simpson's rule, calculate an estimate of the volume of the solid formed, giving your answer to three decimal places.

      ii Calculate the volume using integration, giving your answer to three decimal places.

8 $I = \int_1^5 \ln x\,dx$.

    a Giving your answer to three significant figures, find an approximate value for I:
      i using Simpson's rule with four strips of equal width,
      ii using the mid-ordinate rule with four strips of equal width.

    b Evaluate I using integration by parts, giving your answer to four significant figures.

SKILLS CHECK **6B EXTRA** is on the CD

Examination practice 6: Numerical methods

1 $f(x) = x^3 + \dfrac{2}{x} - 4$, $x \neq 0$.

    a Show that $f(x) = 0$ has a solution in the interval $0.5 \leq x \leq 0.6$.

The solution is to be estimated using the iterative formula $x_{n+1} = \dfrac{2}{4 - x_n^3}$, with $x_1 = 0.52$.

    b Calculate the values of x_2, x_3 and x_4, giving your answers to four significant figures.

    c Using a change of sign method over a suitable interval, show that the solution of $f(x) = 0$ is 0.5180 correct to four significant figures.

2 $f(x) = 0.1x - \ln(x + 2)$, $x > -2$.

    a Rearrange the equation $f(x) = 0$ into the form $x = e^{ax} + b$, stating the values of the constants a and b.

    b Use the iterative formula $x_{n+1} = e^{ax_n} + b$, with $x_1 = -1.1$ and your values of a and b, to find x_2, x_3 and x_4.

    c Hence write down an approximation to the negative root of the equation $f(x) = 0$.

 3 The diagram shows part of the curve with equation $y = f(x)$, where $f(x) = 3x^2 + e^{-x}$. The curve has a minimum at the point A.

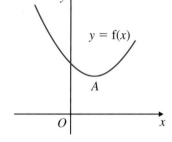

    a Find $f'(x)$.

    b Show that the x-coordinate of A lies in the interval $0.1 < x < 0.2$.

A more accurate estimate of the x-coordinate of A is made using the iterative formula

$$x_{n+1} = \tfrac{1}{6} e^{-x_n}$$

with $x_1 = 0.1$.

    c Write down the values of x_2, x_3, x_4 and x_5, giving the value of x_5 to three decimal places.

4 A sequence is defined by

$$x_{n+1} = \sqrt{(x_n + 12)}, \quad x_1 = 2.$$

    a Find the values of x_2, x_3 and x_4, giving your answers to 3 decimal places.

    b Given that the limit of the sequence is L:

      i show that L must satisfy the equation $L^2 - L - 12 = 0$;

      ii find the value of L.

    c The graphs of $y = \sqrt{(x + 12)}$ and $y = x$ are sketched below.

    On a copy of the sketch, draw a cobweb or staircase diagram to show how convergence takes place.

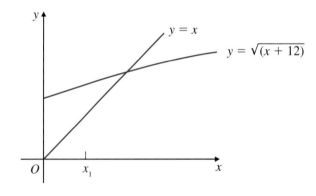

[AQA Jan 2004]

5 **a** By taking logarithms, solve the equation $0.6 = 2^{-x}$, giving your answer to three significant figures.

    b A sequence is defined by $x_{n+1} = 2^{-x_n}$, $x_1 = 0.6$.

      i Find the values of x_2, x_3, x_4 and x_5, giving your answers to three decimal places.

      ii Given that the sequence converges to α, write down an equation in x for which α is a root.

[AQA June 2001]

6 a Solve the equation $3^x = 7$, giving your answer to three significant figures.

   b i Sketch the graphs of $y = 3^x$ and $y = 7 - x^2$ on the same axes.

     ii Hence state the number of roots of the equation $3^x = 7 - x^2$.

   c i Show that the equation $3^x = 7 - x^2$ can be written in the form

$$x = \frac{\ln(7 - x^2)}{\ln 3}.$$

     ii Use the iterative formula

$$x_{n+1} = \frac{\ln(7 - x_n^2)}{\ln 3}, \quad x_1 = 1.5$$

to find the values of x_2 and x_3, giving your answers to three decimal places.

[AQA Jan 2005]

7 Use Simpson's rule with five ordinates (four equal strips) to find an approximation to the integral

$$\int_0^2 \ln(x^2 + 1)\,\mathrm{d}x$$

giving your answer to three decimal places.

[AQA June 2003]

 8 a Use Simpson's rule with five ordinates (four strips) to find an approximation to the integral

$$I = \int_0^1 x \cos x\,\mathrm{d}x,$$

giving your answer to five decimal places.

   b Use integration by parts to find the value of I, giving your answer to five decimal places.

[AQA Jan 2002]

9

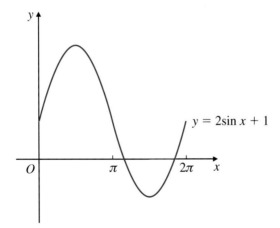

The diagram shows the graph of $y = 2 \sin x + 1, 0 \leqslant x \leqslant 2\pi$.

   a Describe a sequence of transformations by which the graph of $y = 2 \sin x + 1$ can be obtained from the graph of $y = \sin x$.

   b Use the mid-ordinate rule with five strips of equal width to find an estimate of the area enclosed by the curve $y = 2 \sin x + 1$, the x-axis and the line $x = \pi$.

   c Find the exact area by integrating.

Practice exam paper

Answer **all** questions.

Time allowed: 1 hour 30 minutes

A calculator is **allowed** in this paper.

1 a Find $\dfrac{dy}{dx}$ when $y = \dfrac{x}{\tan x}$. *(3 marks)*

   b i Find $\dfrac{dy}{dx}$ when $y = (x^3 + 5)^6$. *(2 marks)*

      ii Hence, or otherwise, find $\displaystyle\int x^2(x^3 + 5)^5 \, dx$. *(3 marks)*

2 a Sketch on the same axes, for $-1 \leqslant x \leqslant 1$, the graphs of $y = \frac{1}{2}(x + 1)$ and $y = \sin^{-1} x$. *(3 marks)*

   b The x-coordinate of the point where the two graphs cross is α. Show that
$0.7 < \alpha < 0.8$. *(3 marks)*

3 A curve has equation $x = y^2 - 4y + \cos y$. Find the value of $\dfrac{dy}{dx}$ at the point $(1, 0)$. *(4 marks)*

4 a Show that the equation

      $\cot^2 x = 2 \operatorname{cosec} x + 7$

   can be written in the form

      $\operatorname{cosec}^2 x - 2 \operatorname{cosec} x - 8 = 0$ *(2 marks)*

   b Hence, or otherwise, solve the equation

      $\cot^2 x = 2 \operatorname{cosec} x + 7$

   giving all values of x in radians, to three significant figures, in the interval $0 < x < 2\pi$. *(6 marks)*

5 A curve C is defined for $0 \leqslant x \leqslant \pi$ by the equation $y = 4 + \sin 2x$.

   a i Find $\dfrac{d^2y}{dx^2}$. *(2 marks)*

      ii Verify that C has a stationary point where $x = \dfrac{3\pi}{4}$ and determine its nature. *(2 marks)*

   b Describe a sequence of two geometrical transformations that maps the graph of
$y = \sin x$ onto the graph of $y = 4 + \sin 2x$. *(4 marks)*

   c Find the area of the region bounded by the curve C, the coordinate axes and the line
$x = \dfrac{3\pi}{4}$, giving your answer in terms of π. *(4 marks)*

6 The curve with equation $y = 4e^x + 6e^{-x} - 11$ intersects the positive x-axis at the point A and
intersects the negative x-axis at the point B.

   a Find the x-coordinates of A and B giving your answers in an exact form. *(6 marks)*

   b Show that the equation of the tangent at A is $y = 5(x - \ln 2)$. *(3 marks)*

   c The tangents at A and B intersect at the point P. Find the x-coordinate of P, giving
your answer in the form $\frac{1}{2} \ln k$, where k is a constant to be determined. *(4 marks)*

7 The diagram shows part of the curve with equation $y = \ln x$.

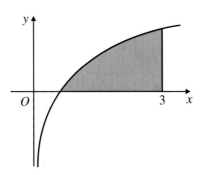

The region bounded by the curve $y = \ln x$, the x-axis and the line $x = 3$ is shaded.

a Use integration by parts to find $\int \ln x \, dx$. *(3 marks)*

b Find the area of the shaded region, giving your answer in a simplified exact form. *(3 marks)*

c The shaded region is rotated through 360° about the x-axis. Use Simpson's Rule with five ordinates (four strips) to find an approximation for the volume of the solid generated. Give your answer to four significant figures. *(5 marks)*

8 The function f is defined for real values of x by $f(x) = 2x - 1$.

a i Find $f^{-1}(x)$, where f^{-1} is the inverse of f and state the domain of f^{-1}. *(3 marks)*

   ii Sketch the graph of $y = |f(x)|$, indicating the coordinates of the points where the graph meets the coordinate axes. *(3 marks)*

   iii Solve the equation $|f(x)| = f^{-1}(x)$. *(3 marks)*

b The function g is defined for real values of x by $g(x) = \dfrac{1}{x^2 + 2}$.

   i Find the range of g. *(1 marks)*

   ii State, with a reason, whether the function g has an inverse. *(1 marks)*

   iii Find gf(x) in the form $\dfrac{1}{ax^2 + bx + c}$. *(2 marks)*

Answers

1 a one–one function   **b** not a function
  c one–one function   **d** many–one function

2 a $f(x) \in \{2.5, 3, 3.5, 4\}$ **b** $f(x) \in \{\frac{1}{4}, \frac{1}{3}, \frac{1}{2}, 1\}$
  c $f(x) \geqslant 0$ **d** $f(x) \geqslant -2$

3 a i See CD   **ii** $f(x) \in \mathbb{R}$
  b i See CD   **ii** $-1 \leqslant g(x) \leqslant 1$
  c i See CD   **ii** $f(x) > 0$
  d i See CD   **ii** $h(x) \geqslant -9$

4 $4\sqrt{3} \leqslant f(x) \leqslant 8$

5 a -13 **b** 4   **c** 12   **d** 7   **e** $3\frac{1}{4}$ **f** 6

6 a 2^{3x+2} **b** x **c** $\dfrac{3}{x}+2$ **d** $\dfrac{1}{3x+2}$

7 a -4.5 **b** -4 **c** $\pm 2\sqrt{2}$ **d** -3 or -6

8 a no inverse, many–one   **b** inverse, one–one

9 a i $f^{-1}(x) = \dfrac{x-5}{2}, x \in \mathbb{R}, f^{-1}(x) \in \mathbb{R}$ **ii** See CD
  b i $f^{-1}(x) = 3 - 4x, x \in \mathbb{R}, f^{-1}(x) \in \mathbb{R}$ **ii** See CD
  c i $f^{-1}(x) = \sqrt{x}, x \geqslant 0, f^{-1}(x) \geqslant 0$ **ii** See CD
  d i $f^{-1}(x) = x^2 + 3, 0 \leqslant x \leqslant 3, 3 \leqslant f^{-1}(x) \leqslant 12$ **ii** See CD

10 a $\frac{17}{19}$ **b** $\frac{1}{2}$ or -1 **c** $\frac{1}{3}$ or 1

11 a $f^{-1}(x) = \dfrac{2}{3-x}, x \neq 3$ **b** 1 or 2

12 a $\frac{1}{4}$ or 1   **b** $f^{-1}(x) = \dfrac{5x-1}{4x}, x \neq 0$ **c** -2

1 a See CD for graph; $(0, 3), (\frac{3}{2}, 0)$
  b See CD for graph; $(0, 1), (\frac{1}{2}, 0)$
  c See CD for graph; $(0, 4)$

2 a i See CD for graph; $(-4, 0), (-2, -4), (0, 0)$
    ii See CD for graph; $(-4, 0), (-2, 4), (0, 0)$
  b i See CD for graph
    ii See CD for graph
  c i See CD for graph; $(0, 1), (\frac{1}{2}\pi, 0), (\pi, -1), (\frac{3}{2}\pi, 0), (2\pi, 1)$
    ii See CD for graph; $(0, 1), (\frac{1}{2}\pi, 0), (\pi, 1), (\frac{3}{2}\pi, 0), (2\pi, 1)$

3 a See CD for graph; $(a, 0), (0, a)$
  b See CD for graph; $(-\frac{a}{2}, 0), (0, a)$
  c See CD for graph; $(0, 2a)$

4 a i See CD for graph   **ii** $\frac{2}{3}$ or 4   **iii** $x < \frac{2}{3}, x > 4$
  b i See CD for graph   **ii** 6 or $\frac{4}{5}$ **iii** $x < \frac{4}{5}, x > 6$
  c i See CD for graph   **ii** $2a$ or $\frac{2a}{3}$ **iii** $\frac{2a}{3} < x < 2a$

5 a $-7, 1$ **b** $-7 < x < 1$ **c** $x \leqslant -7, x \geqslant 1$

6 $\frac{3}{5} < x < 9$

7 a Stretch in the y-direction by scale factor $\frac{1}{2}$, translation by -3 units in the y-direction
  b Reflection in the y-axis, stretch in the y-direction by scale factor 0.4
  c Translation by 2 units in the x-direction, stretch in the y-direction by scale factor 5

8 a See CD for graph; $(3, 5), (5, 2)$
  b See CD for graph; $(0, 1), (1, 0)$
  c See CD for graph; $(-2, 0), (0, -3), (2, 0)$

9 a $y = 3\tan x - 2$ **b** $y = -\sin \frac{1}{3}x$

10 a See CD for graph
  b See CD for graph; $(\frac{1}{4}\pi, 0), (\frac{3}{4}\pi, 1), (\frac{5}{4}\pi, 0), (\frac{7}{4}\pi, -1)$
  c $-1 \leqslant g(x) \leqslant 1$

1 a 6, 9   **b** See CD   **c** $5 \leqslant g(x) \leqslant 9$
  d No, the function is not one–one
  e $x^4 + 10x^2 + 30$

2 a i $-2, 1$ **ii** See CD for graph; $(0, -2), (4, 0)$
  b $-2 \leqslant f(x) \leqslant 1$
  c i $f^{-1}(x) = (x+2)^2$ **ii** $-2 \leqslant x \leqslant 1$
    iii See CD

3 a i $(x+3)^2 + 2$ **ii** $x = -3$ **iii** $f(x) \geqslant 2$
  b $x > -2$ or $x < -4$
  c i Translation by -2 units in the x-direction
    ii Stretch by scale factor $\frac{1}{3}$ in the x-direction and by scale factor 4 in the y-direction

4 a G_2, not one–one
  b i 4, 4.99   **ii** $4 \leqslant f(x) < 5$ **iii** $f^{-1}(x) = 1 - \dfrac{1}{x-5}$

5 a i $fg(x) = \sqrt{x-1}, gf(x) = \sqrt{x} - 1$ **ii** They both equal zero
  b i Translation by 1 unit in the x-direction   **ii** $0 \leqslant h(x) \leqslant 2$
    iii Domain $0 \leqslant x \leqslant 2$; range $1 \leqslant h^{-1}(x) \leqslant 5$ **iv** $h^{-1}(x) = x^2 + 1$

6 a See CD   **b** See CD   **c ii** $(-\frac{1}{2}, 4), (\frac{1}{2}, 4)$

7 a $5a$ **b** See CD   **c** $0, -2a$

8 a See CD for graph; $(2, 0), (0, 4)$
  b i $\frac{4}{3}, 4$ **ii** $x < \frac{4}{3}, x > 4$ **c** 2

9 a i See CD for graph; $(2, 0), (-1, 0), (0, 2)$
    ii No intersection with $y = 4$
  b $f(x) \leqslant 3$ **c** $x = \frac{4}{3}, x = -2; x < -2, x > \frac{4}{3}$

10 a $A_1(-0.5, 0), B_1(0, 3)$ **b** $A_1(-1, 2), B_1(0, 1)$ **c** $A_1(0, -1), B_1(1, 0)$

1 Answers to the nearest degree:
  a $63°$ **b** $-42°$ **c** $70°$ **d** $102°$

2 a 0.64^c (2 d.p.)   **b** -1.04^c (2 d.p.)
  c 1.21^c (2 d.p.)   **d** 2.67^c (2 d.p.)

3 a $23.6°$ (1 d.p.), $156.4°$ (1 d.p.)   **b** $150°, 330°$
  c $24.1°$ (1 d.p.), $155.9°$ (1 d.p.), $204.1°$ (1 d.p.), $335.9°$ (1 d.p.)
  d $60°, 120°, 240°, 300°$

4 a 2.38^c (3 s.f.)
  b 4.71^c (3 s.f.)
  c 3.39^c (3 s.f.), 6.03^c (3 s.f.)
  d 0.785^c (3 s.f.), 2.36^c (3 s.f.), 3.93^c (3 s.f.), 5.50^c (3 s.f.),

5 See CD for PowerPoint solution

6 $60°, 180°, 300°$

7 $-330°, -210°, 30°, 150°$

8 0.52^c (2 d.p.), 1.57^c (2 d.p.), 2.62^c (2 d.p.), 4.71^c (2 d.p.)

9 -2.36^c (2 d.p.), 0.79^c (2 d.p.)

10 a Translate by $-90°$ in the x-direction, reflect in the x-axis
  b See CD
  c They are the same curve

11 a Stretch by scale factor 2 in the y-direction, translate by 1 unit in the y-direction.
  b $(\frac{1}{2}\pi, 3)$ **c** $f(x) \geqslant 3$

12 a Stretch by scale factor $\frac{1}{2}$ in the x-direction, stretch by scale factor 3 in the y-direction
  b $(180, 3)$

13 a $y = \frac{1}{2}\pi, y = -\frac{1}{2}\pi$
  b Reflection in the x-axis, translation by 1 unit in the y-direction
  c $y = 1 + \frac{1}{2}\pi, y = 1 - \frac{1}{2}\pi$

1 b $70.5°$ (1 d.p.), $104.5°$ (1 d.p.), $255.5°$ (1 d.p.), $289.5°$ (1 d.p.)

2 $60°, 109.5°$ (1 d.p.), $250.5°$ (1 d.p.), $300°$

3 0.59^c (2 d.p.), 2.36^c (2 d.p.), 3.73^c (2 d.p.), 5.50^c (2 d.p.)

4 Answers to the nearest degree, where appropriate: $30°, 150°, 194°, 346°$

5 Stretch by scale factor 2 in the x-direction, translation by 1 unit in the y-direction

6 a $60°, 120°, 240°, 300°$ **ii** $30°, 60°, 120°, 150°$

7 LHS $= \dfrac{1}{\sin^2\theta} \times \cos^2\theta = \cot^2\theta = \operatorname{cosec}^2\theta - 1 =$ RHS

8 a $y = \frac{1}{2}\pi, y = -\frac{1}{2}\pi$
  b i $y = 2\tan^{-1}x$ **ii** $y = \pi, y = -\pi$

9 0.32^c (2 d.p.), 2.03^c (2 d.p.), 3.46^c (2 d.p.), 5.18^c (2 d.p.)

1 $x = 3\ln 2$
2 $x = \ln 2$
3 $x = \frac{5}{2}$

4 $x = \dfrac{e^{0.5} - 3}{4}$

5 e^2, e^3

6 a See CD for graph; $(1, 0)$, asymptote $x = 0$

   b See CD for graph; $(3 + e^{-2}, 0)$, asymptote $x = 3$

7 a i $f^{-1}(x) = \ln \dfrac{x}{3}$ **ii** See CD for graph; $(0, 3)$, $(3, 0)$

     iii Range of f: $f(x) > 0$; domain of f^{-1}: $x > 0$,
       range of f^{-1}: $f^{-1}(x) \in \mathbb{R}$

   b i $f^{-1}(x) = \frac{1}{2}e^x$ **ii** See CD for graph; $(\frac{1}{2}, 0)$, $(0, \frac{1}{2})$

     iii Range of f: $f(x) \in \mathbb{R}$; domain of f^{-1}: $x \in \mathbb{R}$,
       range of f^{-1}: $f^{-1}(x) > 0$

8 a $f^{-1}(x) = e^{4x} - 1$ **b** $x \in \mathbb{R}, f^{-1}(x) > -1$ **c** $e^2 - 1$

9 a $f^{-1}(x) = \dfrac{\ln x - 1}{2}$

   b See CD for graph; asymptotes $x = 0$, $y = 0$

   c Range of f: $f(x) > 0$; domain of f^{-1}: $x > 0$, range of f^{-1}: $f^{-1}(x) \in \mathbb{R}$

10 a Translation by -1 unit in the x-direction, reflection in the x-axis, translation by 2 units in the y-direction

   b Stretch by scale factor $\frac{1}{2}$ in the x-direction, stretch by scale factor 3 in the y-direction

Exam practice 3 (page 34)

1 a $x = \ln 4$ **b** $\ln 2 - 2$

2 a $y = \dfrac{e}{e - 1}$ **b** $x = \frac{1}{2}e^3 + 2$

3 a See CD   **b** $(-4, 0), (0, 8), (0, 2)$

4 a See CD for graph; $(0, e - 2), (\ln 2 - 1, 0)$

   b $f^{-1}(x) = \ln(x + 2) - 1$ **c** $x > -2, f^{-1}(x) \in \mathbb{R}$

5 a See CD for graph   **b** $(2e^{-k}, 0)$

6 a Stretch by scale factor $\frac{1}{3}$ in the x-direction

   b Translation by 3 units in the x-direction or stretch by e^{-3} units in the y-direction

   c Reflection in the line $y = x$

7 a Reflection in the y-axis, stretch by scale factor 2 in the y-direction.

   b i $0 < f(x) \leqslant 2$ **ii** Domain $0 < x \leqslant 2$, range $f^{-1}(x) \geqslant 0$

     iii $f^{-1}(x) = \ln 2 - \ln x$ **iv** Yes, since $f(\ln 2) = 1$ and f is decreasing

8 a i Reflection in the line $y = x$ **ii** See CD

   b i Stretch by scale factor 3 in the y-direction   **ii** $f^{-1}(x) = e^{\frac{1}{3}x}$

9 a $p(3) = 27 - 18 - 15 + 6 = 0$

   b $(x - 3)(x - 1)(x + 2)$

   c $y = 0, y = \ln 3$

SKILLS CHECK 4A (page 42)

1 a $5e^x$ **b** $\dfrac{3}{x}$ **c** $-4 \sin x - \cos x$

   d $6x^2 - e^x$ **e** $-\dfrac{2}{x}$ **f** $\frac{1}{4} \sec^2 x$

2 a $-6x \sin(x^2)$ **b** $-6 \sin x \cos x$ **c** $3(2x - 5)e^{x^2 - 5x}$

   d $\dfrac{8}{2x - 5}$ **e** $6 \tan 3x \sec^2 3x$ **c** $-e^{-x}$

3 a $y = \ln 5 + \ln x - \frac{1}{3}\ln(2x + 7)$ **b** $\dfrac{dy}{dx} = \dfrac{1}{x} - \dfrac{2}{3(2x + 7)}$

4 a $20(5x + 6)^3$ **b** $\dfrac{1}{\sqrt{2x - 1}}$ **c** $-\dfrac{1}{(x + 2)^2}$

5 $y + 6x + 11 = 0$

6 1.11^c (2 d.p.)   **b** maximum

7 $(2, 4 - 8 \ln 2)$

8 a $3e^x - \dfrac{2}{x}$

   b $y = 3ex - 2x + 2 - 2 \ln 5$

   c $2 - \ln 25$

9 $(0, 0)$

10 a $e^{3x} - 5e^x + 3e^{2x} - 15$

   b $3e^{3x} - 5e^x + 6e^{2x}$

   c 16

11 a $e^x + e^{-2x}$ **b** -1

12 a $\dfrac{3}{(1 - t)^2}$ **b** $\dfrac{8}{2t + 3}$

13 a $f'(2) = 3.38\ldots > 0$ so f is increasing

   b $f'(2) = -1.25 < 0$ so f is decreasing

14 $\dfrac{dy}{dx} = 3 \cos 3x$, $\dfrac{d^2y}{dx^2} = -9 \sin 3x = -9y$, result follows

15 a $(\frac{1}{3} \ln 2, 2 - 2 \ln 2)$ **b i** $9e^{3x}$ **ii** Minimum

SKILLS CHECK 4B (page 47)

1 a $x^2(x + 4)(5x + 12)$ **b** $3x \sec^2 3x + \tan 3x$

   c $2e^{2x}x^3(2 + x)$ **d** $\dfrac{x}{2x + 6} + \ln \sqrt{x + 3}$

   e $e^{3x}(\cos x + 3 \sin x)$ **f** $2x \cos 2x + \sin 2x$

2 a $\dfrac{3x(x - 6)}{(x - 3)^2}$ **b** $\dfrac{1}{1 - \sin x}$

   c $\dfrac{e^{\frac{1}{2}x}(x - 6)}{4x^4}$ **d** $\dfrac{1 - \ln(x + 1)}{(x + 1)^2}$

   e $-\dfrac{7}{(3x - 2)^2}$ **f** $\dfrac{x \sin x + \cos x}{x^2}$

3 a $(x^2 + 3)^2(5x - 4)^4(55x^2 - 24x + 75)$ **b** $e^x \cos^2 x(\cos x - 3 \sin x)$

   c $\dfrac{4e^x}{(2e^x + 1)^2}$ **d** $\dfrac{2(x \cos 2x - \sin 2x)}{x^3}$

   e $x(1 + 2 \ln x)$ **f** $\dfrac{x(2 - x)}{e^x}$

4 $-e^{-3}$

5 a $(-1, -2e^{-1})$ **b** $(0, 0), (3, 27e^{-3})$

6 b $\dfrac{18}{(x - 3)^3}$

7 $4x + 2y = 5$

8 $\frac{1}{2}$

10 $\dfrac{d}{dx}\left(\dfrac{\sin x}{\cos x}\right) = \dfrac{\cos x \cos x - \sin x(-\sin x)}{(\cos x)^2} = \dfrac{\cos^2 x + \sin^2 x}{\cos^2 x}$

                $= \dfrac{1}{\cos^2 x} = \sec^2 x$

11 a $\cos x\, e^{\sin x}$ **b** $e^{\sin x}(\cos^2 x - \sin x)$

SKILLS CHECK 4C (page 49)

1 $\dfrac{1}{y}$

2 a $\dfrac{1}{\cos y}$ **b** $\dfrac{1}{\sqrt{1 - x^2}}$

3 $\dfrac{(y - 1)^2}{y^2 - 2y - 1}$

4 $\dfrac{1}{4x^{\frac{3}{4}}}$

5 a $\dfrac{3}{y}$ **b** $2x + 6y = 7$

6 $\dfrac{3}{1 - \ln 3}$

7 a $y = x + 1$, $2y = x + 4$ **b** $(2, 3)$

8 $12y = 13x - 25$

9 a $x = \dfrac{1}{\ln 2} \ln y$, $a = \dfrac{1}{\ln 2}$

Exam practice 4 (page 49)

1 a $\dfrac{dy}{dx} = \dfrac{1}{x} - 3$, $\dfrac{d^2y}{dx^2} = -\dfrac{1}{x^2}$

   b $x = \frac{1}{3}$

   c -9, maximum

2 a $p = \frac{1}{2} \ln 5$, $q = -\frac{4}{3}$ **b** $9y + 6x + 8 = 0$

3 a $-3(x^2 + 5x + 4) \sin 3x + (2x + 5) \cos 3x$ **b** $y = 5x + 4$

4 a $\dfrac{2 \sin x - 2x \cos x}{\sin^2 x}$ **b ii** $y = -\frac{1}{2}x + \frac{5}{4}\pi$

5 a $-1, 5$ **b** $9y = x - 2$

6 a i $\dfrac{dy}{dx} = 2x - 3 + \dfrac{1}{x}$

   b ii $x = \frac{1}{2}, 1$ **iii** $\dfrac{d^2y}{dx^2} = 2 - \dfrac{1}{x^2}$ **iv** $x = \frac{1}{2}, \dfrac{d^2y}{dx^2} = -2$; $x = 1, \dfrac{d^2y}{dx^2} = 1$

7 a $\dfrac{x^2 + 6x + 7}{(x+3)^2}$ **b** $x = -7$

8 a $e^x\left(\dfrac{1}{x} + \ln x\right)$ **b** e

9 a -3 **b** $-9(1 + \tan^2 3x)$

10 a i See CD **ii** $f^{-1}(x) = \frac{1}{3}\tan^{-1} x$

   b i $\frac{1}{3}\cos^2 3y$ **ii** $\frac{1}{12}$

11 $3\cos x - 3\cos^3 x$

12 a $-\dfrac{6}{(2x-1)^4}$ **b** $-3\sin x\, e^{3\cos x}$ **c** $2e^{2x}$ or $2y$

13 a $-\frac{3}{5}$ **b** -1, maximum

SKILLS CHECK 5A (page 61)

1 a $\frac{1}{3}e^{3x+1} + c$ **b** $e^{-u} + c$ **c** $-\frac{1}{2}e^{-2t} + c$

   d $\frac{1}{3}\ln|x+1| + c$ **e** $\frac{2}{5}\ln|1 + 5x| + c$ **f** $-\frac{1}{2}\ln|1 - 6x| + c$

   g $3\sin\frac{1}{3}x + c$ **h** $-\frac{1}{4}\cos 2y + c$ **i** $\frac{1}{9}\sin 3x + c$

2 2

3 $y = \frac{1}{3}e^{3x} - x^2 + \frac{2}{3}$

5 a 3.5 **b** 9 **c** 3

6 a $[\ln|x-4|]_5^8 = \ln 4 - 0 = \ln 2^2 = 2\ln 2$ **b** 0.75

8 $a = -7, b = 2$

9 $\frac{1}{2}e^{x^2+2} + c$

10 a $3\sin^{-1}\left(\dfrac{x}{3}\right) + c$ **b** $-6\sqrt{9-x} + c$

11 0.785 (3 s.f.)

12 $-\ln 1 + e^{-x} + c$

13 $-2\ln|1 + \cos x| + c$

14 a $\frac{1}{10}(2x-3)^{\frac{5}{2}} + \frac{1}{2}(2x-3)^{\frac{3}{2}} + c$ **b** $\frac{1}{8}(2+x)^8 - \frac{2}{7}(2+x)^7 + c$

15 a $-\frac{1}{3}\cos^3 x + c$ **b** $\frac{1}{3}$

16 a i $25\left(\left(\frac{1}{15}\right)^2 - x^2\right)$ **ii** $\frac{1}{5}\sin^{-1}(5x) + c$

   b i $25\left(\left(\frac{3}{5}\right)^2 + x^2\right)$ **ii** $\frac{1}{15}\tan^{-1}\left(\dfrac{5x}{3}\right) + c$

17 a $\frac{1}{2}\ln(4 + x^2) + \frac{1}{2}\tan^{-1}\left(\dfrac{x}{2}\right) + c$ **b** $\sin^{-1} x + 3\sqrt{1 - x^2} + c$

SKILLS CHECK 5B (page 65)

1 a $(2x+3)\sin x + 2\cos x + c$ **b** $\frac{1}{2}x\sin(2x+3) + \frac{1}{4}\cos(2x+3) + c$

2 a $\frac{1}{3}xe^{3x} - \frac{1}{9}e^{3x} + c$ **b** $\frac{1}{3}xe^{3x} - \frac{4}{9}e^{3x} + c$

3 5.93 (3 s.f.)

4 a $-\frac{5}{2}x\cos 2x + \frac{5}{4}\sin 2x + c$

5 a $-\frac{1}{2}xe^{-2x} - \frac{1}{4}e^{-2x} + c$ **b** $\frac{3}{4}e^{-2} - \frac{5}{4}e^{-4}$

6 a -2 **b** $\pi^2 - 4$

7 a $-\frac{3}{4}e^{-2} + \frac{1}{4}$

8 $2 - \dfrac{5}{e}$

SKILLS CHECK 5C (page 68)

1 a $\ln 2$ **b** $\frac{1}{2}\pi$

2 $\frac{1}{2}\pi(e^2 - 1)$

3 $\frac{93}{5}\pi$

4 $\pi\ln 2$ **b** $\frac{5}{6}\pi$

5 a $A(0, 1), B(1, 2)$ **b** $\frac{7}{15}\pi$

6 a $(3, 3)$ **b** See CD

7 $\frac{3}{10}\pi$

8 a $\cap$ shaped quadratic curve through $(-3, 0)$, $(0, 9)$ and $(3, 0)$

   b $\frac{226}{5}\pi$ **c** $\frac{81}{2}\pi$

9 a i 1 **ii** $\frac{1}{4}\pi^2 - 2$

   b $\pi(\frac{1}{4}\pi^2 - 2)$

10 a $-e^{-2x}(\frac{1}{2}x + \frac{1}{4}) + c$

11 a $x\tan x + \ln|\cos x| + c$

Exam practice 5 (page 70)

2 a i $\dfrac{1}{x}$ **ii** $-\dfrac{1}{x^2}$

3 a $6x(x+1)$ **b** $k = \frac{1}{6}$

4 $\frac{1}{2}e^2 + \frac{1}{6}$

5 $\frac{1}{4}$

6 a i See CD for graph; $(0, 1)$, $(\frac{1}{2}\ln 2, 0)$

    ii See CD for graph; $(0, -4)$, $(\frac{1}{2}\ln 5, 0)$

   b i $\frac{1}{2}e^{2x} - 2x + c$

7 $\frac{2}{3}$

8 $\pi - 2$

9 a $\dfrac{3}{2\sqrt{3x+4}}$ **b** 2

10 $a = 1, b = 2, c = 2$; $1.5 - 2\ln 2$

11 a $\frac{1}{4}x^4\ln x - \frac{1}{16}x^4 + c$ **b** $\frac{232}{5}$

12 $\frac{13}{81}$

13 a $-\frac{1}{5}e^{3-5x} + c$ **b** $-\frac{1}{3}|\ln 4 - 3x| + c$

14 a i $\frac{1}{2}\ln(9 + x^2) + c$ **ii** $\frac{1}{3}\tan^{-1}\left(\dfrac{x}{3}\right) + c$

   b $a = \frac{10}{9}, b = \frac{1}{3}$

15 a $\frac{6}{5}\pi$ **b** $2y = 5x - 1$

   c i $3x\sqrt{x^2 + 3}$ **ii** $\frac{1}{3}(8 - 3^{\frac{3}{2}})$

16 a $1 + \dfrac{12}{3x+2} + \dfrac{36}{(3x+2)^2}$

   b i $\frac{1}{3}\ln|3x+2| + c$ **ii** $-\frac{1}{3}(3x+2)^{-1} + c$

   c 37.8 (3 s.f.)

17 a $\ln|1 + \sin x| + c$ **b** $-\dfrac{1}{1 + \sin x} + c$

18 $\frac{1}{2}(3 + \ln x)^2 + c$

SKILLS CHECK 6A (page 78)

Note that, where appropriate, you must make a comment about the change of sign.

1 a $f(5) = -0.290..., f(6) = 0.817...$

   b $f(-1) = -1.41..., f(0) = 1$

   c $f(4.1) = -0.277..., f(4.2) = 0.439...$

   d $f(-4) = -1.17..., f(-3) = 0.092...$

   e $f(1.2) = 0.105..., f(1.3) = -0.427...$

2 a See CD

   b 2

   c $f(2) = 0.693..., f(3) = -3.90...$

3 a See CD **c** $n = 1$

4 a ii $1.5275..., 1.5196..., 1.5218..., 1.521$

   b ii $-0.338295..., -0.339630..., -0.339680..., -0.33968$

   c ii $0.2116..., 0.2143..., 0.2150..., 0.215$

   d ii $1.3234..., 1.3240..., 1.3241..., 1.324$

   e ii $7.5078..., 7.5070..., 7.5067..., 7.507$

5 c 1.217 (4 s.f.)

6 a 2.28 (3 s.f.) **b** See CD **c** $e^x - 2x - 5 = 0$

7 a $f(1) = 2.54..., f(2) = -1.41...$

   b $1.6944..., 1.6960..., 1.696$

8 a See CD

   d $2.28462..., 2.28128..., 2.28056...$

   e 2.28 (2 d.p.)

9 a $f(5) = -3, f(6) = 4$

   b ii $4.25, 2.5156..., -0.4179..., -1.9563..., -1.0431...$; sequence appears to be converging, but not to the root in the interval $5 < x < 6$

   c ii $5.2915..., 5.4005..., 5.4407..., 5.4555..., 5.4609...$; sequence appears to be converging to a root in the interval $5 < x < 6$

   d 5.46 (2 d.p.)

   e Exact root $= 2 + 2\sqrt{3} = 5.46410...$, so values agree to 2 d.p.

SKILLS CHECK 6B (page 84)

1 0.366 (3 d.p.)

2 a 126 (3 s.f.)

   b $\frac{1}{3}(e^6 - 1)$

   c Approximately 6%

3 4.51 (3 s.f.)

4   16.566 (3 d.p.)

5   **a**   17.354 (3 d.p.)

       b   Use more strips, ensuring that an even number is chosen

6   **a**   See Section 2.1

       b   0.4388 (4 d.p.)

       c   Very close, so it is a good approximation

7   **a**   0.5405 (4 d.p.)

       b   **i**   1.698 (3 d.p.)   **ii**   1.698 (3 d.p.)

8   **a**   **i**   4.04 (3 s.f.)

          ii   4.08 (3 s.f.)

       b $5 \ln 5 - 4 = 4.047$ (4 s.f.)

Exam practice 6 (page 85)

1   **a** $f(0.5) = 0.125$, $f(0.6) = -0.450\ldots$

       b   0.5182, 0.5180, 0.5180

       c $f(0.51795) = 0.000328\ldots$, $f(0.51805) = -0.000336\ldots$

2   **a** $a = 0.1$, $b = -2$

       b $-1.104165\ldots$, $-1.104538\ldots$, $-1.104572\ldots$

       c -1.105 (3 d.p.)

3   **a** $6x - e^{-x}$

       b $f'(0.1) = -0.304\ldots$, $f'(0.2) = 0.381\ldots$

       c $0.1508\ldots$, $0.1433\ldots$, $0.1444\ldots$, 0.144

4   **a**   3.742, 3.968, 3.996   **b**   **ii** $L = 4$ **c**   See CD

5   **a**   0.737 (3 s.f.)   **b**   0.660, 0.633, 0.645, 0.640   **c** $x = 2^{-x}$

6   **a**   1.77 (3 s.f.)   **b**   **i**   See CD   **ii**   2

       c   **ii**   1.418 (3 d.p.), 1.463 (3 d.p.)

7   1.434 (3 d.p.)

8   **a**   0.38182 (5 d.p.)   **b**   0.38177 (5 d.p.)

9   **a**   Stretch by scale factor 2 in the y-direction, translation by 1 unit in the y-direction

       b   7.21 (3 s.f.)

       c $4 + \pi$

Practice exam paper (page 88)

1   **a** $\dfrac{\tan x - x \sec^2 x}{\tan^2 x}$ or $\cot x - x \operatorname{cosec}^2 x$

       b   **i** $18x^2(x^3 + 5)^5$ **ii** $\dfrac{1}{18}(x^3 + 5)^6 + c$

2   **a**   See CD

3 $-\frac{1}{4}$

4   **b**   0.253, 2.89, 3.67. 5.76

5   **a**   **i** $-4 \sin 2x$

          ii   Minimum

       b   Stretch in x-direction, by scale factor $\frac{1}{2}$; translation by $\begin{bmatrix} 0 \\ 4 \end{bmatrix}$

       c $3\pi + \frac{1}{2}$

6   **a** $x_A = \ln 2$, $x_B = \ln \frac{3}{4}$

       c $x_P = \frac{1}{2} \ln \frac{3}{2}$

7   **a** $x \ln x - x + c$

       b $3 \ln 3 - 2$

       c   3.238 (4 s.f.)

8   **a**   **i** $f^{-1}(x) = \frac{1}{2}(x + 1)$, domain: all real values

          ii   See CD

          iii $x = \frac{1}{5}$, 1

       b   **i** $0 < g(x) \leq \frac{1}{2}$

          ii   g has no inverse as g is not a one-one mapping

          iii $gf(x) = \dfrac{1}{4x^2 - 4x + 3}$

SINGLE USER LICENCE AGREEMENT FOR CORE 3 FOR AQA CD-ROM
IMPORTANT: READ CAREFULLY

WARNING: BY OPENING THE PACKAGE YOU AGREE TO BE BOUND BY THE TERMS OF THE LICENCE AGREEMENT BELOW.

This is a legally binding agreement between You (the user or purchaser) and Pearson Education Limited. By retaining this licence, any software media or accompanying written materials or carrying out any of the permitted activities You agree to be bound by the terms of the licence agreement below.

If You do not agree to these terms then promptly return the entire publication (this licence and all software, written materials, packaging and any other components received with it) with Your sales receipt to Your supplier for a full refund.

YOU ARE PERMITTED TO:

- Use (load into temporary memory or permanent storage) a single copy of the software on only one computer at a time. If this computer is linked to a network then the software may only be used in a manner such that it is not accessible to other machines on the network.

- Transfer the software from one computer to another provided that you only use it on one computer at a time.

- Print a single copy of any PDF file from the CD-ROM for the sole use of the user.

YOU MAY NOT:

- Rent or lease the software or any part of the publication.

- Copy any part of the documentation, except where specifically indicated otherwise.

- Make copies of the software, other than for backup purposes.

- Reverse engineer, decompile or disassemble the software.

- Use the software on more than one computer at a time.

- Install the software on any networked computer in a way that could allow access to it from more than one machine on the network.

- Use the software in any way not specified above without the prior written consent of Pearson Education Limited.

- Print off multiple copies of any PDF file.

ONE COPY ONLY

This licence is for a single user copy of the software

PEARSON EDUCATION LIMITED RESERVES THE RIGHT TO TERMINATE THIS LICENCE BY WRITTEN NOTICE AND TO TAKE ACTION TO RECOVER ANY DAMAGES SUFFERED BY PEARSON EDUCATION LIMITED IF YOU BREACH ANY PROVISION OF THIS AGREEMENT.

Pearson Education Limited and/or its licensors own the software.
You only own the disk on which the software is supplied.

Pearson Education Limited warrants that the diskette or CD-ROM on which the software is supplied is free from defects in materials and workmanship under normal use for ninety (90) days from the date You receive it. This warranty is limited to You and is not transferable. Pearson Education Limited does not warrant that the functions of the software meet Your requirements or that the media is compatible with any computer system on which it is used or that the operation of the software will be unlimited or error free.

You assume responsibility for selecting the software to achieve Your intended results and for the installation of, the use of and the results obtained from the software. The entire liability of Pearson Education Limited and its suppliers and your only remedy shall be replacement free of charge of the components that do not meet this warranty.

This limited warranty is void if any damage has resulted from accident, abuse, misapplication, service or modification by someone other than Pearson Education Limited. In no event shall Pearson Education Limited or its suppliers be liable for any damages whatsoever arising out of installation of the software, even if advised of the possibility of such damages. Pearson Education Limited will not be liable for any loss or damage of any nature suffered by any party as a result of reliance upon or reproduction of or any errors in the content of the publication.

Pearson Education Limited does not limit its liability for death or personal injury caused by its negligence.

This licence agreement shall be governed by and interpreted and construed in accordance with English law.